# INTERDISCIPLINARY DESIGN IN PRACTICE

# INTERDISCIPLINARY DESIGN IN PRACTICE

Edited by
ROBIN SPENCE, SEBASTIAN MACMILLAN and PAUL KIRBY

Thomas Telford

Published by Thomas Telford Publishing, Thomas Telford Ltd, 1 Heron Quay, London E14 4JD. URL http://www.thomastelford.com

Distributors for Thomas Telford books are
*USA*: ASCE Press, 1801 Alexander Bell Drive, Reston, VA 20191-4400, USA
*Japan*: Maruzen Co. Ltd, Book Department, 3–10 Nihonbashi 2-chome, Chuo-ku, Tokyo 103
*Australia*: DA Books and Journals, 648 Whitehorse Road, Mitcham 3132, Victoria

First published 2001

A catalogue record for this book is available from the British Library

ISBN: 0 7277 3008 8



Typeset by Ian Kingston Editorial Services, Nottingham, UK.
Printed and bound in Great Britain by MPG Books, Bodmin.

# Contents

# Foreword

Effective building construction or assembly tends to depend more on the quality of the junctions between components than on the components themselves. As Sir Alan Cockshaw observes in his contribution to this book, the same is true of the different professions and industries in the built environment. 'We have world class architects and engineering designers and very able construction and manufacturing industries.... We are collectively capable of giving far better value for money than we do.... The problem is not the quality of the individual components but the way they come together.'

It is this concern that links the authors of this book, brought together here under an interdisciplinary editorial team from Cambridge University's Departments of Architecture and Engineering. Sir Michael Latham and Sir John Egan's reports on the UK construction industry have both brought the issue of collaboration into the public arena – but others were grasping the opportunities offered by interdisciplinary collaboration long before these two reports. And it is the lessons learnt and the conclusions drawn from such long experience that are perhaps the most invaluable aspect of this book.

One of the most intriguing of these insights is the fact that specialisation is unlikely to decline because the alternative, generalisation, never matches the results achieved by effective collaboration between skilled specialists. Such collaboration depends on excellence in our own fields, an understanding of the concerns of our fellow participants and an inclusive, shared approach to the task in hand. The ability to listen to society, to our clients and to each other and to communicate our enthusiasm – or 'passion' as Sam Price describes it – is all-important.

Interdisciplinarity can be explored in many modes – but, within the development and construction industries, it is interdisciplinary *design* that holds the key. For, as Sir Jack Zunz, quoting Sir Alan Muir Wood, points out, 'Design is the central element of architecture – and of engineering'. It is also the primary influence both on the entire process of construction and commissioning and on the effectiveness and enjoyment of the constructed work. Effective *interdisciplinary* design has to be the way ahead as, to quote Sir Alan Cockshaw again, 'What we can do together that we cannot do apart is truly amazing'.

*Peter Carolin CBE*
*Emeritus Professor or Architecture, University of Cambridge*

# Introduction

*Sebastian Macmillan, Robin Spence and Paul Kirby*

Why is interdisciplinary design in the built environment important? After all, architects and engineers in Britain are renowned for their ability to deliver innovative and exciting buildings. Aren't they already performing well? Such a perception tells only part of the story. While at its best the construction industry is recognized as excellent, government task forces on both urban design and construction have suggested the need for a step change in the way we design and deliver the built environment. With the support of several government initiatives to promote innovation, the whole construction sector is being expected to raise the quality of its products and the efficiency with which it delivers them. Design is central to the production of the built environment, and designers themselves are not immune to the need to embrace change. As much as every other part of the industry, designers need to review their business processes and identify what improvements are possible. This book is offered as a contribution to the debate. It suggests that interdisciplinary working is one of the most important improvements that the design professions need to address. It describes the experiences and achievements of leading design practitioners, together with pioneering educational initiatives, and reports the findings of several research projects into interdisciplinary design processes.

Recommendations for the design of the built environment can be traced back 2000 years to Vitruvius' Ten Books on Architecture (Morgan, 1960), in which he proposed six fundamental principles – order, arrangement, eurythmy, symmetry, propriety, and economy. As well as the qualities that buildings and cities should exhibit, he also identified the qualities necessary in those who were to be responsible for design – including a knowledge of history, philosophy and music. Vitruvius' books were addressed to the Emperor and sought to influence the design of public buildings.

Circumstances today are not so very different. Indeed, they are remarkably similar, as in the UK we face a national campaign to achieve better public buildings (Department for Culture, Media and Sport, 2000). Nor is this campaign entirely new. More than ten years ago, the Prince of Wales entered the public debate with his *Vision of Britain* (1989), in which he proposed 'Ten Principles we can build upon'. Later, the Royal Fine Art Commission tried to identify *What makes a good building?* (RFAC, 1994) and subsequently, the Department of Environment launched its *Quality in Town and Country* initiative (DoE, 1995). Research was commissioned into the quality of urban design, as reported by Rowley (1998), which identified 50 desirable attributes, grouped into four bundles: functional and social use considerations; natural environment and sustainability considerations; visual considerations; and the urban experience.

Following the election of the new government in 1997, the design of the built environment and the efficiency of the construction industry became a renewed focus of attention at senior government levels. The Deputy Prime Minister commissioned both a Construction Task Force, led by Sir John Egan, and an Urban Design Task Force, led by Lord Rogers. The Construction Task Force report *Rethinking Construction* (DETR, 1998) made a number of recommendations for improving the quality and efficiency of UK construction, with design subsequently identified by the Deputy Prime Minister as a vital component:

> *Good design is an integral and essential part of 'Rethinking Construction'. It is a key to many of the performance targets; it is a key to reducing construction time and defects; it is a key to sustainability and to respect for the environment. In the broadest sense it is the key to respect for people, whether they be users of a building or passers by... (Rt Hon John Prescott, 1999).*

Similar observations were made by the Urban Task Force about the central importance of design (1999):

> *Design is a core problem-solving activity that not only determines the quality of the built environment – the buildings, public spaces, landscape and infrastructure – but also delivers many of the instruments for the implementation of an urban renaissance (Urban Task Force, 1999).*

The Treasury Task Force (2000) went further in defining the contribution of design:

> *At its broadest, design is the process in which intelligence and creativity are applied to a project in order to achieve an efficient and elegant solution. ... good design is not an 'optional extra',*

> *rather it is inherent in the way the brief is responded to from the very beginning. Design encompasses functional efficiency, structural integrity, sustainability, lifetime costing, and flexibility as well as responsiveness to the site and to its setting. ... Good design involves creativity, and it should lead to simplification and to savings in cost. ... it can increase outputs and add to the quality of service. It can also give the facility a competitive advantage in attracting both customers and staff. Good design can also contribute to wider policy objectives, such as those relating to the protection of the environment. Good design ... adds value in the following ways: functionality; reducing whole life costs; service enhancement; architectural quality and wider social and environmental benefits (Treasury Task Force, 2000).*

These sentiments were subsequently picked up, endorsed and expanded by the government's Better Public Buildings Group. In the foreword to the group's publication *Better Public Buildings*, the Prime Minister commits the government to making a step change in the quality of building design in the public sector (Department for Culture, Media and Sport, 2000).

What conditions are necessary to achieve such a step change in the design of the built environment? One criticism of the design process as currently practised is made by the government's own Construction Best Practice Programme. Under the heading 'Integrating design and construction' its Web site states:

> *The urge to reinvent is difficult to contain. It is becoming increasingly understood that there is waste in the design process with many design decisions taken earlier in the process needing to be reworked by later players. The extended and disconnected supply chain, brought about to some extent by contracts, does not encourage dialogue between the client's designers and the suppliers' designers, contracted to the main contractor.*

Dialogue, together with improved understanding and mutual respect among the parties, are beginning to emerge as crucial. As Nick Raynsford, Minister for London, Planning and Construction, said in 1999:

> *High quality building and urban design can be achieved consistently only by a team approach; clients, professional advisors, contractors, and specialist contractors working together in equality and mutual respect. Design is integral to the whole construction process and part of the responsibility of the whole construction*

> *team. ... Achieving good design in all these respects is a real challenge and ... requires the combined talents of the whole supply chain. (Raynsford, 1999).*

These are the conditions that, in this book, we take as representing interdisciplinary design. They embrace first an inclusive definition of the design team – covering all those involved, from the inception of a project to its final hand-over; second, mutual respect among the parties, in which the contributions and skills of each party, though varying in scale, type or extent, are acknowledged as necessary to the whole enterprise; third, an awareness among all the parties of the total scope of the project, and a shared focus on the successful delivery of the completed product; and finally a willingness among the parties to communicate ideas and proposals across discipline boundaries.

Such sentiments are not entirely new. In 1965 in a manifesto for improving quality in building design, Ove Arup proposed that:

> *successful teamwork requires that each member of the team understands what the others are doing and respects them, and that they are united in a common purpose: to produce good architecture. (Arup, 1985).*

Both Sir Michael Latham (1994) and Sir John Egan (DETR, 1998) expressed similar views – that there was a need to foster better teamwork and to reduce the adversarial nature of the building design and procurement processes. They considered such changes, among others, essential as the basis for a more competitive and mature industry, focused more directly on the needs of its clients and users.

But although the notions of teamwork, collaboration and interdisciplinary working are not new – indeed they are concepts to which many now subscribe – they continue to be hard to realize in practice. It is important to ask why the building industry has been so prone to conflict and litigation, and why design professionals from different disciplines find it so hard to work together smoothly and effectively. These issues were considered in the Latham and Egan reports, and a number of improvements in the operations of design and construction teams were recommended. Key to the problems identified by Latham was the fragmentation of the industry and the need for new contractual relationships to enhance collaboration in providing clients with what they want. Egan went further, and called for a revolution in the way the construction industry operates, creating an integrated project process fostered by long-term partnerships among builders, designers and suppliers in place of one-off teams.

Both of these reports also recognize that changes in the education, training and professional formation of the entire team are essential to bring about such changes, requiring new courses and new broader curricula on the part of professional courses and industry training programmes alike; a need anticipated by the Ove Arup Foundation in hosting the Madingley Seminar on Education for the Built Environment in 1991 (Cambridge Programme for Industry, 1991). At this meeting a new model for interdisciplinary education was proposed by Cooper and Stonehouse (1991):

> *It is quite clear that in order to improve the education of professionals for the built environment we must improve the way in which the various disciplines work together ... . The model we propose starts from the basis of the separate disciplines but promotes greater relationships between the disciplines as the education progresses ... . It leads to a breaking down of the narrow professional definitions of the disciplines and a growth of interdisciplinary knowledge and skills.*

Many of the changes proposed in these documents of the 1990s have already begun to happen. But, beyond such changes, it may also be that collaboration has been difficult to achieve because of a lack of good models and practical guidance about how to achieve it. What are the principles, what should be the procedures, how should we expect the process and the outcome to be different from normal practice? A fundamental aim of this book is to help to make good this deficiency by offering some current models and hard-won experience of what it takes to make teamwork effective. We believe it is the first book to attempt this.

Many of its chapters were first presented in the form of talks to the students on the Master's Course on Interdisciplinary Design for the Built Environment. This course was launched at Cambridge University in 1994, partly as a response to the Madingley Seminar, and with the clearly defined aim of promoting better understanding among the different professions involved in the design team. During the last six years, more than 100 designers from different professions – architects; civil, structural and services engineers; surveyors; project managers and others – have been working together on this course, in a joint endeavour to understand the task of design from each other's perspectives; as well as to improve their understanding of the broader context of design – economics, the natural landscape, urban planning, environmental sustainability, professional roles and responsibilities.

During this period the course has attracted many distinguished speakers to Cambridge, and two informal Conferences have been held, which have generated some stimulating new ideas, and at which some remarkable case studies in interdisciplinary design have been presented. Another outcome of the IDBE initiative has been the generation of a new programme of research into the behaviour of interdisciplinary design teams, conducted in collaboration with several leading design firms.

Much of this material is ground-breaking, and together it sets out a radical new view of the design team, how it can be organized, what its aims should be, and what it can deliver in terms of excellence in design. The editors believe that the sum of this material, if more widely known and understood, could contribute to bringing about the desirable but elusive changes in the culture of the construction industry so often proposed in the past.

The intention of this book is to edit and reframe some of the best of this thinking into a form which will offer, for design professionals and others concerned about achieving quality and value in our built environment, an introductory perspective on interdisciplinary design, and propose some ways forward. It contains contributions from many prominent individuals representing different professional viewpoints, among them architects Richard Saxon, Ian Ritchie and Dean Hawkes, engineers Sir Alan Cockshaw, Michael Dickson, Sam Price and Sir Jack Zunz, and developer Peter Rogers. Case studies provide illustrations and examples. All of these contributions were originally prepared as verbal presentations; they have been rewritten and updated for inclusion as chapters in this book. We have tried nevertheless to preserve the freshness of their personal presentation style.

The book also presents and reviews recent innovative experiences of education for interdisciplinary design both in the university and practice environments. And it includes summaries of best practice in the design process drawn from management studies and academic research.

The book is organized in three parts. In Part 1, six chapters by practitioners present the problems facing the design team from different perspectives, and illustrate particular solutions to those problems through description of case study buildings.

It is well understood that any general change in the fragmented culture of the design team will only come about through changes in the education and processes of professional formation in the different design professions. Part 2 considers the task of education, at the level of the school, at undergraduate level, and through postgraduate courses both in universities and in the workplace, and describes some innovative approaches to the education of designers at all these levels.

Part 3 shows what has been learnt about interdisciplinary design through some recent research studies that have examined closely the workings of design teams. By drawing both on these observational studies and on the management science literature on teamwork, this part of the book offers research-based good practice guidance.

More about the contents of each part is found in the reviews which introduce them.

## References

Arup, O. (1985) The problem of producing quality in building. Talk to the Westminster Chamber of Commerce, 27 April 1965, reprinted in *The Arup Journal*, Spring 1985.

DETR (1998) *Rethinking Construction: the Report of the Construction Task Force*, July 1998.

Cambridge Programme for Industry (1991) *Education for the Built Environment*. University of Cambridge.

Cooper, P. and Stonehouse, R. (1991) Education, Research and Training, in *Education for the Built Environment* (ed. Cambridge Programme for Industry). University of Cambridge.

Department of Culture, Media and Sport (2000) *Better Public Buildings: a proud legacy for the future*. DCMS.

Department of the Environment (1995) *Quality in Town and Country*. DoE.

Latham, M. (1994) *Constructing the Team. Final Report of the Government/ Industry Review of Procurement and Contractual Arrangements in the UK Construction Industry*. HMSO, July 1994.

Morgan, M. H. (translator) (1960) *Vitrivius – The Ten Books on Architecture*. Dover Publications, London.

Rt Hon John Prescott (1999) DETR Press release, 19 July 1999.

Prince of Wales (1989) *A Vision of Britain*. Doubleday, London.

Raynsford, N. (1999) DETR Press release, 19 October 1999.

Rowley, A. (1998) Quality of urban design: a study of the involvement of private property decision-makers in urban design. *Journal of Urban Design*, **3**(2).

Royal Fine Art Commission (1994) *What Makes a Good Building?* RFAC.

Treasury Task Force (2000) *How to Achieve Design Quality in PFI projects*. Technical Note 7, HM Treasury.

Urban Task Force (1999) *Towards an Urban Renaissance*. E. & F. N. Spon, London.

# Contributors

**Sir Alan Cockshaw**

Sir Alan Cockshaw graduated in civil engineering from the University of Leeds, and spent his early career in both the public and private sectors. He joined Fairclough Civil Engineering in 1973 and was appointed Chief Executive 1978, becoming a member of the main board of the Fairclough Construction Group in 1981. In 1982 Fairclough acquired the Press Group and in so doing created the AMEC Group, where Sir Alan became Group Chief Executive in 1984. He was Chairman from 1988–1997 during a period of huge international expansion. In October 1998 he was appointed by the UK Government as Chairman of English Partnerships. Sir Alan is also the founding Chairman of PCS International Limited, and a member of the boards of several other companies. He has been involved in major company formations, acquisitions and disposals in many parts of the world. He is also a past president of the Institution of Civil Engineers.

**Michael Dickson**

Michael Dickson is an engineer who studied mechanical sciences at Cambridge University followed by structural engineering and town planning at Cornell University, USA. He has worked on a wide range of major building engineering and urban regeneration projects such as headquarters buildings for IBM and British Airways in England, headquarters for RWE Essen in Germany, a new main line station for Stuttgart 21, a large urban development in the centre of Shanghai, and the Japanese Pavilion in Hanover. Michael is Founding Partner and Chairman of Buro Happold Consulting Engineers. He was recently Chairman of the IStructE Working Group 'Building for a Sustainable Future – Construction without Depletion' and a member of Lord Justice Taylor's Working Group on the Safety of Sports Grounds. He is a Director of the Theatre Royal, Bath, and is Chairman of the Construction Industry Council.

**Christopher French**

Chris French has worked in the public sector for all of his professional life, first as a project architect designing public buildings, including social housing, schools, social services and health buildings and culminating in the design and construction of the William De Ferrers Community School in Essex, and latterly as a managing architect. He served as Essex County Council's Principal Development Architect for 10 years, responsible for a small multi-disciplinary team carrying out research and development through a series of live projects. During this time he completed his doctorate into primary school design from 1970–1995. He is now responsible for the County Council's design standards, managing a small group who commission and supervise the authority's panel of multi-disciplinary consultants working on around 250 projects worth £30 million a year. He serves on several national working groups advising the government and industry on school design.

**David Gann**

David Gann holds the IMI/Royal Academy of Engineering Chair in Innovative Manufacturing and leads the Programme on Innovation in the Built Environment at SPRU, University of Sussex. His main areas of research are on innovation in the built environment and the management of technical change in project-based firms. He was Advisor to the UK Deputy Prime Minister's Construction Taskforce and is a Board member of The Housing Forum and a Member of the EPSRC's Advisory Panel. He has worked extensively with government departments in the UK and in other countries and has provided advice to industry, and to agencies such as the OECD on policies for innovation and technical development. He publishes widely in academic journals and the media, and his latest books are *Building Innovation: Complex Constructs in a Changing World* and *Digital Futures: Making Homes Smarter*.

**Dean Hawkes**

Dean Hawkes has been Professor of Architectural Design at the Welsh School of Architecture since 1995. He received his initial architectural training at the Regional College of Art, Manchester, and studied for his PhD at Cambridge. From 1970 to 1995 he was University Lecturer in Architecture at Cambridge, and Director of the Martin Centre for Architectural and Urban Studies between 1980 and 1989. His research interests are in environmental studies in architecture and the relationship between architectural science and theory and design. He publishes widely and his most recent book is *The Environmental Tradition: Studies in the Architecture of Environment*. His architectural work has been published and exhibited internationally. He was a finalist in the 1998 competition for the design of the Welsh National Assembly building in Cardiff and also in the international competition for *A House for the Future* promoted by the National Museum of Wales in 1999.

**Paul Kirby**

Paul Kirby graduated with a degree in Architectural Engineering from the University of Leeds. Between 1976 and 1993 he worked as a building services consulting engineer, first with Max Fordham and Partners and latterly with Whitby and Bird, accumulating a wide ranging experience of design and design teams. At its inception in 1993, he joined the team running the Interdisciplinary Design for the Built Environment Masters Programme at the University of Cambridge. Currently the Course Director, he also teaches in the University's Departments of Engineering and Architecture and contributes to research projects. He is a Fellow of Wolfson College.

**Sebastian Macmillan**

Sebastian Macmillan trained as an architect at Liverpool University followed by a PhD at the Royal College of Art. He is a member of the Royal Institute of British Architects, and ran an architectural practice for ten years. In 1984 he set up Eclipse, a consultancy specializing in research on the design and management of the built environment. He has written many publications for the government's Energy Efficiency Best Practice programme, and edited the technical manual *Environmental Management for Hotels*. He has a part-time appointment at Cambridge University's Martin Centre, where he is responsible for coordinating the Centre's research programme. In 1999 he was appointed Sector Programme Manager for Construction by the Engineering and Physical Sciences Research Council. He is active in various construction sector improvement initiatives, including Technology Foresight and the Construction Research and Innovation Strategy Panel.

**Andy Parker**

Andy Parker has over 15 years experience in the design engineering and construction of complex manufacturing facilities. Having held a variety of posts within AMEC, including Design Team Leader, Project Manager and Head of Mechanical Services, he is currently leading the development of the company's Pharma-Chem business in Europe. He graduated in Civil Engineering from Loughborough University and has a Masters Degree from Cambridge University in Interdisciplinary Design for the Built Environment. He has a particular interest in the benefits of integrated multi-disciplinary design teams. Andy is a member of the Chartered Institute of Building Services Engineers' Executive Board. He also sits on the Edge Committee and the Editorial Panel of the *Building Services Journal*.

**Sam Price**

Sam Price is a structural engineer and joint founding partner of the practice of consulting engineers Price & Myers. He is a graduate of Trinity College Cambridge, where he studied Mechanical Sciences and was a rowing blue. He worked for Ove Arup & Partners as an engineer from 1960, joining the multi-disciplinary practice of Arup Associates in 1967 as a project engineer for a series of major schemes including the Lloyds Administration Building by the Medway at Chatham. With Robert Myers he founded the practice of Price & Myers in 1978. Recent noteworthy projects include university colleges in both Oxford and Cambridge with architects MacCormac Jamieson & Pritchard and van Heyningen & Haward. He is a Fellow of the Institution of Civil Engineers and an Honorary Fellow of the Royal Institute of British Architects.

**Ian Ritchie**

Ian Ritchie studied architecture in Liverpool and London. In 1981 he established Ian Ritchie Architects, and in the same year co-founded the design engineering practice of Rice Francis Ritchie in Paris. His current positions include Royal Academician, Chairman of the Royal Academy of Arts Collection & Library Committee, Chairman of Europan UK, Commissioner at the Commission for Architecture and the Built Environment (CABE) and DTI IBIS Project Gatekeeper. He is the recipient of many national (RFAC, RIBA, Civic Trust, Structural Steel, Concrete Society etc.) and international awards, including the Académie d'Architecture Silver Medal (2000), the International Association of Bridge and Structural Engineers (IABSE) 'Outstanding Structure Award' (2000) and the Commonwealth Association of Architects Robert Matthew Award for Innovation and the Advancement of Architecture (1994). He has published a number of books including *(well) Connected Architecture*, *The Biggest Glass Palace in the World*, and *Ian Ritchie, Technoecology*.

**Peter Rogers**

Following extensive experience with major construction contractors in the UK and internationally, Peter Rogers joined Stanhope in 1985 as a main board director. Initially responsible for the construction of the £700m Broadgate development, he subsequently took on responsibility for the construction activity across all the company's projects. He has since been accountable for the procurement of several billion pounds of commercial, arts and historic buildings. He has acted as an advisor to a variety of clients including the Royal Opera House, the Ministry of Defence, and the Tate Modern on projects that include theatres, art galleries and the £500m White City retail scheme. He has also assessed and monitored lottery-funded projects for the Arts Council and the National Heritage Memorial Fund. He has contributed to numerous construction industry initiatives including Technology Foresight. He is a Trustee of the Victoria and Albert Museum and an advisor to the Royal Shakespeare Company.

**Ammon Salter**

Ammon Salter has a D.Phil. in Science and Technology Policy Studies from the Science and Technology Policy Research, University of Sussex. He has worked as a consultant and advisor on a number of innovation projects, including the Program on Globalization and Regional Innovation Systems (Toronto University) and research on Ontario's Innovation Systems. He was involved in a study of the Economic Benefits of Basic Research for the UK Treasury. He is currently engaged in an Engineering and Physical Sciences Research Council-funded project contributing to a major initiative on mapping, measuring and managing innovation in project-based organizations; this involves working closely with six firms to develop better knowledge about the management of technology and how technical capabilities are developed and enhanced.

**Richard Saxon**

Richard Saxon is chairman of Building Design Partnership, the international architect-engineer practice. He is also chairman of the Reading Construction Forum, a construction industry think tank for customers, consultants and constructors. He is a member of the Treasury Construction Group and the Strategy Groups of the Government Construction Client Panel and Construction Best Practice Programme. He was a founder and past president (1995–96) of the British Council for Offices. His books include *Atrium Buildings* and *The Atrium Comes of Age*. Recently completed design projects include the masterplan for the All England Tennis Club at Wimbledon, the London Headquarters of J P Morgan, and several new stores for Marks & Spencer.

**Robin Spence**

Robin Spence is a structural engineer and Reader in Architectural Engineering in the Department of Architecture at Cambridge University.

After two years of structural design practice with Ove Arup & Partners he spent two years teaching at University of Zambia, followed by two years as Research Officer for the Intermediate Technology Development Group, researching small-scale building materials technology in Africa and India. After obtaining his Ph.D. he has been with the Architecture Department since 1975, and a Director and Joint Director of the Martin Centre since 1985. He was the Director of the IDBE Course from 1993 to 1997. He is also a Director and Chairman of Cambridge Architectural Research Ltd, and a Fellow of Magdalene College. His research interests are in disaster mitigation, construction technology and interdisciplinary design and he has published extensively on these subjects.

**John Steele**

John Steele gained a Masters degree in Architectural Engineering from the University of Leeds before joining AMEC Capital Projects – Construction. In 1998, he was seconded to the Department of Architecture at the University of Cambridge as a research assistant to map interdisciplinary design processes during the early stages of building projects, at the same time registering for a Ph.D. at the Department of Civil and Building Engineering, Loughborough University. After completing his doctorate in 2000, he returned to AMEC as part of the Innovation & Technology team. He was then seconded to Loughborough University as Gatsby Innovation Fellow where he and Professor Simon Austin are developing a management consultancy offering advisory and educational services to organizations involved in complex, multi-disciplinary projects that can benefit from improved processes, planning, management and collaborative working.

**Sir Jack Zunz**

Jack Zunz is a Founding Trustee of Ove Arup Partnership. He joined Arups in 1950 and became successively a Senior Partner, Chairman of Ove Arup & Partners and Co-Chairman of the Ove Arup Partnership in the UK and the Arup group of practices worldwide. He has been involved with many well-known projects in the UK and elsewhere. As Chairman of the Ove Arup Foundation he was centrally involved in several initiatives to improve the education of professionals working in the Built Environment – the IDBE programme at Cambridge University, the Cities programme at the London School of Economics and the Chair for Civil Engineering Design at Imperial College for Science Technology and Medicine. He has received a number of awards, including the Institution of Structural Engineers' Gold Medal. He is the author and co-author of a number of publications. He was knighted in 1989.

# PART ONE

# Interdisciplinary design in practice

# Introduction

## *Robin Spence*

During the last five years, the UK construction industry has entered a period of rapid and irreversible change. New legislation, new Government guidance, new alliances, new working practices and new attitudes have all begun to change the way construction projects are formulated, designed and built.

Interdisciplinary working is at the heart of these changes. Both of the two most influential government reports which have contributed to the process of change (Latham, 1994; DETR, 1998) have urged greater collaboration between disciplines, between professions and between different parts of the supply chain, on an industry previously characterized by debilitating and cost-inflating conflict.

The chapters which follow show that better interdisciplinary collaboration is needed to achieve at least three necessary objectives of change. First, better collaboration is needed in order to enable the industry to improve its performance: to deliver what the client needs at a predictable and reasonable cost, and within an acceptable time. To achieve this, collaboration and understanding is required throughout the entire chain of participants in a project from the client, through the design team and major construction company, to the subcontractors and suppliers of components.

Secondly, better collaboration is needed to understand and control the growing array of technical requirements which any project must meet; and in this respect none is today more important than the goal of sustainability. Sustainability is an elusive concept, but clearly demands from architects, engineers and constructors a deeper understanding of the interaction of the building with its local and the global environment over a period of time which stretches throughout and even beyond the life of the building; and it calls for a holistic view of the design process on the part of all contributors to it.

Thirdly, better collaboration is needed simply to produce good architecture. Good architecture, characterized by Ove Arup as 'a way of building which delights the heart', can only be achieved as a result of an intense and open dialogue between all members of the design team. Its enemy is the kind of professional exclusiveness and elitism that has too often prevailed in the past.

Each of the following chapters is a view on interdisciplinarity offered by a leading practitioner active today in the process of change. Each offers a different perspective on the aims and methods of interdisciplinary working from the point of view of that author's role in the construction industry and attitude to the aims of design. They do not therefore constitute a tidy, unified picture of a single approach or understanding of the state of the industry and where it must try to go. But taken together they form an extraordinarily rich and diverse set of perspectives on the construction industry.

The first three chapters are primarily concerned with the performance of the industry as a whole and its ability to meet the expectations of its clients. In Chapter 1, Richard Saxon, architect and Chairman of BDP, one of the UK's leading interdisciplinary design practices, starts by pointing out that the UK, in spite of having some of the lowest input costs in Europe, has the highest cost of building. He describes the current process of change, and explains what collaboration really involves in changing the nature of contractual relationships throughout the supply chain. He argues that this will inevitably bring dramatic changes to the way that professionals work, with an emphasis on interdisciplinary understanding, flexibility, and the ability to update their skills continuously throughout their professional careers.

The theme of industry performance is taken up again in Chapter 2, The Partnership Approach, by Sir Alan Cockshaw. He makes the case for better collaboration across the whole design and construction team as a force for a more competitive industry which can deliver a better quality product at cost and on time, and he reviews progress towards this aim over the last five years.

In Chapter 3, Peter Rogers of Stanhope Properties Plc looks at the industry from the point of view of the client. He argues that the client takes the biggest risk in committing to a new building project, and must always be at the centre of the process of design; he discusses the alternative approaches to delivery, and argues for a balance in design between the need to innovate and the need to make best use of existing products to keep down cost and construction time. All this has implications for the education of design professionals.

In Chapter 4, Michael Dickson, Chairman of Buro Happold discusses how technological advance in building design demands closer interdisciplinary collaboration in particular between architects and engineers.

Describing and illustrating a series of innovative building projects with which his practice has been involved, he shows how the Bruntland Report notion of sustainable development can be used as the generator of a new engineering architecture with its own aesthetic and impact; this is the result of a true cross-disciplinary collaboration, depending on ever developing technical resources available to designers and greater capability of the construction industry.

Chapter 5, by Ian Ritchie, takes up the themes of sustainable development and the need for closer engineer–architect collaboration for creative design. He argues that sustainable development is one key to achieve a more integrated world; but art is an essential ingredient of all good design; good architecture must therefore be the result of synthetic thinking between the whole design team; inventiveness may come from any part of the design team; and what is important is to create the conditions for true collaboration. Illustrations are drawn from the work of the design firms RFR and Ian Ritchie Associates.

Producing good buildings is also the fundamental aim of the design teams described in Chapter 6 by Sam Price, Partner of Price and Myers, Consulting Engineers. Producing a good building demands that an idea must be pursued with a passion which must be shared by everyone involved. This is why interdisciplinary collaboration is needed from the beginning. Three different models of collaboration are explored along with the architecture which was produced by the process. No one method is ideal; each was suited to the building it produced. But in each case not only the head but also the heart was involved.

## References

DETR (1998) *Rethinking Construction: the Report of the Construction Task Force*, July 1998.

Latham, M. (1994) *Constructing the Team. Final Report of the Government/ Industry Review of Procurement and Contractual Arrangements in the UK Construction Industry*. HMSO, July 1994.

**CHAPTER ONE**

# Changing construction culture

*Richard Saxon*

## Introduction

This chapter looks at the circumstances which led to the Latham Review of 1994 and the Egan Task Force of 1998. The construction industry is being changed by a combination of new business and procurement concepts, collaborative working, new processes for design and construction, information technology and human resource issues.

The Latham Review founded the Construction Industry Board of client, government and industry, whilst the Egan Task Force created the Movement for Innovation. The industry is now set on a path of continuous improvement, focused on client need. The new professional will need continuous development to suit this new culture.

## The road to Latham/Egan

The UK construction industry was in bad shape, but is getting better. We have suffered the highest output prices in Europe with some of the lowest input costs; our clients are often dissatisfied and litigious, making less use of the industry than in other advanced nations; and professionals and contractors are not making enough money, even at these high prices. The only good things to be said are:

- that UK designers and constructors are able to produce some of the best architecture in the world
- that there is underway a deep-seated change campaign which aims to revolutionize the industry.

The British way of designing and constructing buildings is unique and looks odd by international standards. As craftsmanship in traditional materials was replaced in the early twentieth century by knowledge of new materials, so we in the UK progressively de-skilled the main and specialist contractors and vested knowledge and responsibility in university-trained professionals. We developed a cost control system based on standard descriptions of work, robbing contractors of initiative in how things are made. As construction technology progressed, however, this division between the supposedly knowledgeable and responsible and those who actually develop and install systems became illogical and ludicrous. The social, cultural and legal gulf between professionals and trades became a barrier to collaboration.

The introverted concerns of the 400 000 businesses in the UK industry made us lose touch with our customers through decades of public sector dominance. As the private sector took the lead in the 1980s, client dissatisfaction spawned the new diversity of procurement paths, the collapse of professional fee scales, the rise of the project manager, the construction lawyer and the triumph of imported building products with their skilled support. The depression of 1990–93 put the lid on it, crushing the life out of the UK construction firms and professionals and creating a realization that the whole squabbling mess had to be reformed, root and branch. The Conservative Government launched the Latham Review with the inspired realization that unless clients took centre stage, the industry could not change. Constructing the Team (Latham, 1994) gave direction to a mass of initiatives. The incoming Labour Government in 1997 decided to boost the process with a client task force led by Sir John Egan. The report *Rethinking Construction* (DETR, 1998) replaced Latham's call for evolution with a plea for revolution, doing things completely differently. In particular Egan called for an end to price tendering as the method of assembling project teams, replacing it with long-term supply chain relationships driven by measurement to continuous improvement. Thanks to the effect of Latham, the Egan message was taken up rapidly, notably by the government client (HM Treasury, 1999). The new era is taking shape and changing the culture, roles and potential of everyone involved in that new-found phenomenon, the Supply Chain.

## Five forces at work

The Tavistock Institute produced a study for the Government of barriers to change in the industry (Holti and Standing, 1996). It has been very useful in identifying the multi-fold forces for change and giving them some structure. Tavistock identifies five interlocking fields converging to create the comprehensive change:

- business and procurement concepts
- the trend to collaborative working
- design and construction method development
- information technology as a driver and enabler
- human resources issues.

Business and procurement changes are generated by clients more than by the industry itself. Clients, especially the regular ones, have become proactive and have begun to compare how they run their own businesses with how they buy construction. The idea that construction is utterly different from manufacturing or from other service industries because of its one-off, site-based nature has been largely discredited. The last two decades of thinking about business methods is now arriving in construction with a rush. Among the many aspects of this, the Private Finance Initiative stands out. Government now seeks to relate to its property as a charge-paying tenant without responsibilities for upkeep or utilities, concentrating on its core business. The PFI provider has to deliver a long-term supply of managed, serviced, maintained space. Not even the most caring property developer has ever done that in the UK, though it is common in the USA. Adopting it is changing priorities utterly. Consortia are forming of facility managers (a newish but future core profession), designers, financiers and contractors to compete for these projects. Lifetime costing replaces the obsession with initial capital costs. Quality becomes properly valued. Professional designers can even be equity partners to put their interest in the performance of the product up front. They can sit around the consortium table and put quality where it counts. The conventional institutional development on a 25-year repairing lease is now finding PFI style property a serious competitor for the private business tenant.

Collaborative working is a style adapted from manufacturing and service industry. Instead of the single-project, tender-based selection of suppliers to work under an adversarial contract, there is a shift to managed 'supply chains' with a high degree of trust. Just as car companies now buy their windscreen wipers from two sources rather than the open market, and get fully alongside the suppliers to ensure the success of the whole, so partnerships of clients, consultants, contractors and suppliers are appearing (Figure 1.1). For some companies, like Birse, the conversion from carnivorous to herbivorous style has been Damascene. Therapists are brought in to break the old confrontational habits of a lifetime and teach the win–win philosophy. What everyone is seeking is Japanese-style long-termism, coupled with the continuous improvement only committed partners can achieve. It is not cosy cartelism, but sharp, tough love. In the process the alliance-making is restructuring the industry and redefining professionalism again. Suddenly, we are all altruists, concerned for our customers first, both internal and external types.

Design and construction methods are also looking at the manufacturing model and at the near parallel of offshore oil construction. In the

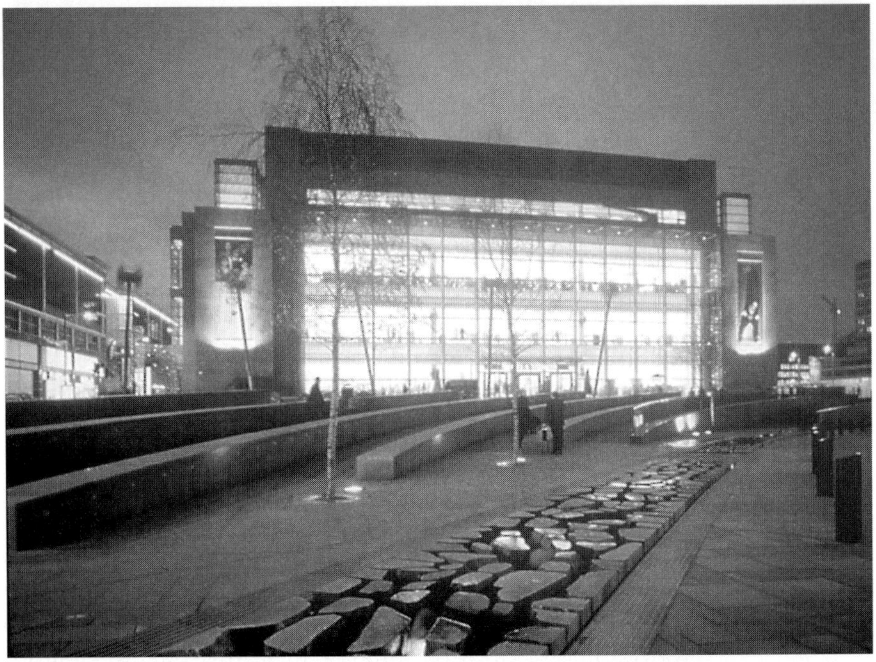

*Figure 1.1 Marks & Spencer Store, Manchester, 1999. The store replaced one destroyed by the IRA bomb in 1996. Marks & Spencer, BDP and Bovis Lend Lease worked closely together to create a large, landmark building and a commercial success in only three years. The long-term relationships between all the parties exemplified the trend to partnerships.*

latter, costs have come down over 30% by collaborative working and concurrent engineering. In the former, Boeing can now build a first-off plane with no faults, thanks to well-managed teamwork and use of intelligent object modelling. Concurrent engineering in construction terms is the reversal of the twentieth century trend to distancing and de-skilling the trade and main contractor. Bringing them in at stages C and D, in RIBA terms, enables greater speed and economy even in a one-off project. In a series of projects it allows progressive process improvement to cut costs deeply and improve quality at the same time. Standardization of specification, knowledge engineering and other IT-based methods are all re-ordering and integrating the design and construction process. The future of design and build, as a procurement path, is seen to lie with a separation of project design work, which should not innovate technically, from continuous, off-line research and development. This is revolutionary stuff. The design–build experiment

'Building Down Barriers', for Defence Estates, demonstrated a lot of the potential.

Information technology is the great enabler of most of the other changes. Five key target concepts are emerging:

- the importance of global, high-speed communication and e-commerce on the Internet
- single project databases, shared by all participants, usually as extranets
- electronic documents, emerging from structured workflow and in managed systems
- knowledge management within firms, with knowledge-based systems as expert advisors and automated designers
- object modelling, where both catalogue items and one-off designs are modelled with all their characteristics and dimensions.

The emerging IT-based industry links client, team and site closely, and enables design sign-offs based on real client understanding through virtual reality presentation and site assembly without surprises. It naturally supports collaborative working, concurrent design and predictable performance.

The designer sitting at a drawing-board sized screen, 'telepresent' wherever needed and with Web access to everything and everyone involved, is not far away. Even now, work styles are changing from the support of paper-based approaches towards something new.

Which brings us to the human factors: Tavistock thinks that these are crucial, limiting the workability of some concepts, encouraging others. People's skills and attitudes are being challenged continuously by all these innovations. Customer focus comes hard to architects; collaborative working seems unnatural to contractors; project team leadership is new to clients; design involvement is a challenge to specialist contractors. Few people over 30 are at home with IT. Most do not like the hours or stress that the industry has demanded.

The workload of re-skilling and attitude reconstruction through training is colossal, and yet training is the weakest aspect of the industry and its clients. It can't just be done through the education of the next generation; it has to be done with the current practitioners or they will be retired early, perhaps very early.

## The reform process today

The process in 2000 has run a substantial way from its launch in the early 1990s. The Construction Industry Board has been succeeded by

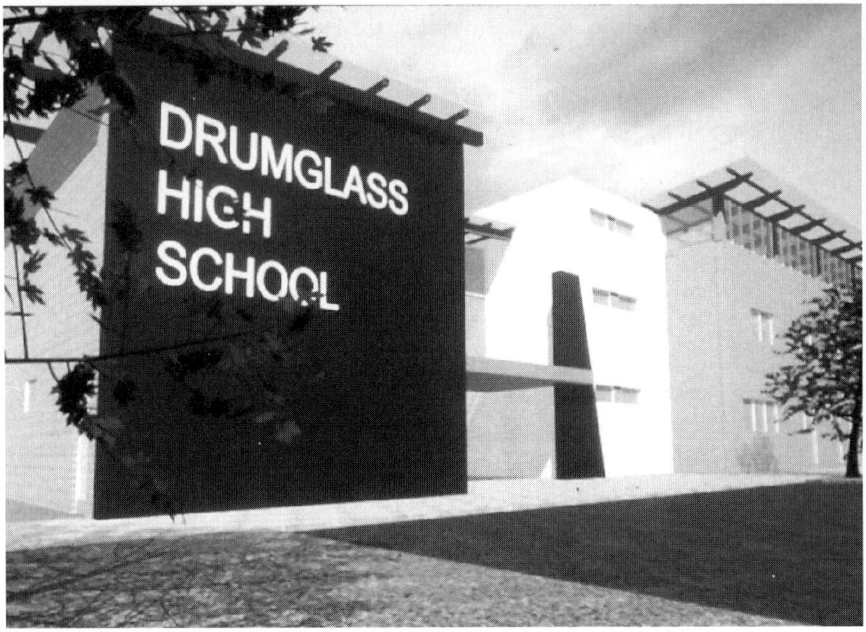


*Figure 1.2 Drumglass School, Northern Ireland, 2000. This modest secondary school is highly innovative in its procurement. BDP acted as PFI provider, retaining a contractor and facilities manager and in turn being retained as designer under a design–build contract. The small team produced a rapid performance, completing the building before parallel PFI projects had reached commercial close. This is an M4I demonstration project.*

the Strategic Forum for Construction, chaired by Sir John Egan. The Movement for Innovation (M4I; see Figure 1.2) has generated billions of pounds worth of demonstration projects showing new methods to all. A Construction Best Practice Programme provides a wealth of advice and knowledge to those responding. The Confederation of Construction Clients has launched a Clients' Charter of good customer practice (Confederation of Construction Clients, 2000) and the government client has committed central and local government to becoming model customers. Key performance indicators are being used to track the industry's progress and for any firm to benchmark itself.

The leading regular clients have formed supply chains which they use regularly and are achieving significantly higher performance than before. R&D activity in groups like the Reading Construction Forum

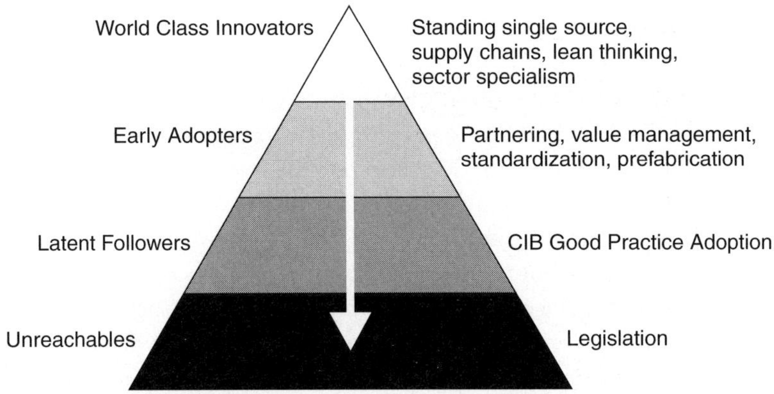


*Figure 1.3 The pyramid of progress. The Construction Industry Board illustrates the way change moves through any population, dividing players into four layers:*

- *Innovators, leading the introduction of changes.*
- *Early Adopters, aware and able to follow once the lead is given.*
- *Latent Followers, unaware but able to follow once the advantages can be made clear to them.*
- *Unreachables, unaware and unable to follow.*

*M4I, the Reading Construction Forum and the Design Build Foundation were set up by Innovators and are filling up with Early Adopters. The Construction Best Practice Programme speaks to Latent Followers and tries to push back the number of unreachables.*

and the Design Build Foundation is setting a pace for the mainstream to follow.

The pace of change is high at the top of the pyramid (Figure 1.3): the pioneers and early adopters. Cultural inertia is, however, enormous. Great resistance is being shown by many customers and contractors towards win–win practice. Consultants are fearful of loss of role and status and many are holding back. Product suppliers are just beginning to see the potential for themselves. There is still a long way to go.

## The new professional

So what are the resultant changes to the professional landscape? I see a business-driven end to the culture of introspection, with the academic art-form of architecture taking its place as a part, but not the core of the discipline. Customer focus will determine what professionals offer and what is seen to add value.

There will be great changes in what engineers, architects, quantity surveyors or builders do with their working hours. Processes are being reassessed and the old, uncoordinated professional plans of work are melting away. New maps are emerging, with clients and specialist contractors on the map.

A concept of the interdisciplinary team is thus widening. Alliances between groups of professionals are extending to all other members of the supply chain and becoming a strategic weapon in competition. Many professionals are working as or for clients and contractors rather than as consultants.

The cyber-professional is emerging as a formidable creature. He or she can work across the world or across the country electronically and can have access to the experience of their whole firm and indeed of the industry. We have to get used, however, to the client eavesdropping on our screen and adding comments and to a more structured, less intuitive work process.

The new professional is forced to take personal development very seriously. Without the interpersonal skills, business skills and technology skills required, we are unemployable. Some skills date fast, so blocks of time and income have to be set aside for training, both by employer and employee. A lot of the productivity gain has to be ploughed back. The original years at university and the original definition of one's discipline will fade in significance quite quickly as the continuous development ethos takes hold. Non-cognate graduate entry into construction will become common.

The Interdisciplinary Design Course no doubt originally was conceived in a relatively traditional way. Its future lies on the broad deck across which lies the whole supply chain. Our challenge is to learn to pull together (Latin: CON-TRACTOR).

## References

Confederation of Construction Clients (2000) *Construction Clients Charter*.

DETR (1998) *Rethinking Construction: the Report of the Construction Task Force*, July 1998.

HM Treasury (1999) *Achieving Excellence*.

Holti, R. and Standing, H. (1996) *Innovations in the Organisation of Construction. What is needed to make them work?* Tavistock Institute interim report to the DoE.

Latham, M. (1994) *Constructing the Team. Final Report of the Government/Industry Review of Procurement and Contractual Arrangements in the UK Construction Industry*. HMSO, July 1994.

CHAPTER TWO

# The partnership approach

*Sir Alan Cockshaw*

## The UK construction industry

We must all accept that the construction industry in the United Kingdom is nowhere near as efficient as it should be and that the important changes and improvements we have seen in the last few years must be continued and accelerated. We have world-class architects and engineering designers and very able construction and manufacturing industries. There is no doubt that we are collectively capable of giving our clients far better value for money than we do at the moment and desirably and unashamedly at the same time improving the profitable performance of our individual businesses. The problem is not the quality of the individual components, but the way in which they come together.

We clearly have the greatest ability to influence the cost of and time to execute a project during the concept stage. Put another way, more quality time spent at the beginning of a project, whether in the development of its design, in its planning or procurement, will have the greatest influence on the quality of the end result.

Whilst much greater efficiency will undoubtedly accrue from better management of all the interfaces within the life cycle of a project, there is no doubt that the way in which we manage the very beginning is where we can create the greatest benefit. The consequences of not preparing properly are well understood. Late changes and variations can have an enormous cost and time impact on any project.

When I started out from university in the structural design of buildings I was often involved, even when working with some outstanding architects, in seeking to determine a structure to fit a design concept which was already well advanced. We were certainly very secondary in the design process. In turn we had very little whatsoever to do with the building services, which were much less complex than in the modern buildings of today. Separatism really was quite the order of the day, and more than a few traces of that methodology live with us even now.

Happily things have moved on a long way and we clearly do have much better integration of the design processes than we ever had before.

Clients have always sought better value for money; that certainly is not new. But today's clients have often been involved in making investments in other parts of the world and they have other reference points with which to compare their UK projects. Global investors such as the international oil companies have offices and laboratories, as well as production facilities. So do those in the pharmaceutical industry, the food industry and many other areas too. They have experienced more efficient ways of working, more certainty:

- certainty that the projects on which they are embarking will be delivered to the budgeted cost or less
- certainty that their project will be completed within the programme period or a shorter one
- certainty that the quality standards required will be achieved.

They accept that safety and environmental issues in the broadest sense are part of the process and not simply something passed on to the contractor who is at the end of the line. They know for example that improved safety performance leads to better projects. They know that they can get certainty in all of these matters elsewhere in the world and if they do not get it here in the UK they can easily invest their capital somewhere else. And of course they will choose overseas companies to design and build their facilities for them rather than UK companies.

## The partnership approach

So how can we define the design team so as to improve the overall service we offer to our clients? It is interesting to look at how other similar industries have tackled the problem. The offshore oil and gas sector has done so with rigour and enthusiasm through the Cost Reduction Initiative for the New Era or CRINE. CRINE was introduced in 1992 and in only a few years produced dramatic savings of 30% in the capital cost of offshore structures. I chaired the first meeting of the government-sponsored group tasked with looking at competitiveness in the broader onshore process plant industry here in the UK, and I have no doubt that similar if not better savings will accrue as this initiative develops.

In both of these schemes, most of the savings are achieved through far more effective partnerships and alliances between clients, designers and constructors. However, we must also recognize that some of the cost savings accrue from a significant reduction in what can best be referred to as 'man-to-man marking' so prevalent in such projects in the past. More than 70% of our business in AMEC Engineering is now in partnering and alliance projects where by working together to closely aligned objectives with a joint commitment at all levels, significant improvements are being achieved not only in capital expenditure but also increasingly in operating expenditure.

Whilst the evolution of alliancing and partnering in the offshore industry grew very quickly in just over two years in the early 1990s, this concept has been very much at the heart of AMEC's approach for many years, especially working closely with clients in the high-tech manufacturing industries.

I am very firmly of the view that in the future successful projects of all kinds will only be so if the relationships between the client, the designer, the contractor and the supply chain are closer than ever before – much more of a partnership. This must not in any way limit the creative thinking required in the evolution of any scheme. It should simply seek to ensure that whatever the concept, it is delivered in the most time-effective and cost-effective way, at the same time as ensuring that quality, safety and environmental standards are maintained throughout.

We must move further and further away from the adversarial approach which has increasingly beset our industry in the UK for such a long time. I expect that trends towards such partnerships will continue, broaden and accelerate into the more traditional areas of the construction industry which have been slowest to change. Government investment, for example, not unnaturally has moved more slowly than any other, based primarily upon the perceived requirements of public accountability – a concept which we would all agree is absolutely necessary in theory but in practice often flawed. If estimates of cost and time are made which when achieved produce an acceptable return, then public accountability is best served by ensuring that those overall parameters are achieved – not by seeking to monitor every individual item of investment throughout the design and construction process and generally over-policing the whole project. This is the fundamental change that international oil companies have recognized and grasped with both hands.

If you pick the right team the client can have all the guarantees that are required for the efficient completion of his project, yet competition can be preserved in the evolution and development of the concept, not in the marginal pricing of the recommended solution. Again this is becoming normal practice in the oil and gas industry.

## Harnessing new technologies

It is now not by any means unusual for AMEC's subsidiary Watson Steel to receive drawings for a steel structure and after careful review to suggest that there are significantly cheaper ways of achieving the same solution, sometimes by preserving the same form and on other occasions by examining a more radical alternative. Regularly this team has been able to offer solutions which create significant savings compared with the prices which will be quoted for the recommended solution; it is often easy, for example, to achieve savings simply by changing the design of connections to more readily mechanize the fabrication process.

Partnerships can of course be infinitely more successful the more we integrate our technological capability. Certainly from AMEC's perspective this is beneficial, since:

- the higher the technology requirements
- the higher the quality standards and
- the larger and more complex the project,

the higher are the barriers to entry, and therefore inevitably competition is progressively reduced.

Modern technology is increasingly allowing interactive interrogation of design concepts, details and of course cost plans. Very sophisticated architectural concepts can be exported electronically by an architect, imported by a manufacturer such as Watson's, developed and then re-exported to the architect and engineer in a digital form. The benefits to the architect and engineer are that they have a high confidence in the transmission of their ideas and, having received the manufacturer's developed information, the great advantage of being able to interrogate the structure and its interface with other building elements.

Poor CAD in inexperienced hands will greatly exacerbate the problem, since the traditional monitoring service usually provided by experienced draughtsmen is no longer present. This is a vital area for future development.

Many of us have found working with Japanese companies very stimulating over the last 10 to 15 years. Our experience has always been on a totally integrated design and construct basis, always with agreed cost plans, always with definitive quality requirements – and always successful. Lots of time spent up front and no time whatsoever necessary at the end of any of them.

## Public–private partnerships

However, I do believe we can go even further in redefining the design team. I refer to the increasing development of partnerships between the public and private sector – both with central government and with local government too. In 1991 in Hulme, Manchester, the Labour leader and I were challenged by Michael Heseltine to set up a partnership to bid for the first round of the City Challenge. Our conditions were that we would only do it if we could:

- look at the whole area not just part of it;
- offer a radical and comprehensive approach; and
- seek to harness the spending power of all the government departments who needed to be involved – as well as the EU.

The City Challenge bid was won and we in AMEC had £7.5m per annum for the next five years to transform one of the most difficult areas of the city – not a lot of money for such a major problem.

Our original vision for Hulme was further developed by a series of individual architectural practices working closely with the community and the partners in the joint venture. We sought funds for housing, for highways and for an extension of the metro system. We created the opportunity for commercial and industrial development and real space for people to enjoy.

In the first three years, over £110 million was raised from the public and private sectors to tear down the worst parts, refurbish others and progressively rebuild an entire community. The process has involved not only the original partners and the government, but the entire community, who have been fundamental to the decision-making process from concept through to completion.

Many of us have 'redefined our design team' to tackle the challenge posed by the larger PFI initiatives. These require fully integrated teams right from the beginning through to the end of the sensible life cycle of the individual projects. Financial engineering is the key. AMEC's first involvement in a direct sense was the Manchester Metro system – enormously successful by any standards.

In AMEC we have had to invest more and more in the best available technology. We have had to learn to use it more effectively in all that we do. We have had to learn to move it more and more effectively around the world. Already in 1995 we could transmit design information electronically straight to machines that cut steel in our offshore yards in Newcastle and in our fabrication works in Bolton. The more we develop our technology, the more we unlock the ever growing potential.

More and more of our clients are looking for total solutions; it is not enough now just to provide a world-class product. Clients investing in the power or water industry simply want to agree a unit price to buy electricity or water. They don't want to buy the turbines, the boiler or the construction. Detailed solutions will increasingly require bigger and stronger teams and therefore 'redefining the design team' will become a continuing process of improvement.

If we don't together redesign the design team into a more proactive integrated forward-looking team, not only will we not give the value for money that our clients so urgently seek, but we will not see the return of sensible profitability to all of us engaged in this marvellous industry. I believe passionately that the skills that exist in British companies, whether in design, manufacture or construction, when working together can create tremendous added value both at home and increasingly overseas. Of course the demand for our services will increase in direct proportion to the improved level of service we offer. What we can do together that we cannot do apart is truly amazing.

## A review of progress

Much of the above was written for a contribution to a discussion at the Royal Society of Arts on 'Redefining the Design Team' in 1995. In the five and a half years since I thought about my contribution to that discussion, much has changed, but nowhere near as much or as fast as I would have wished. In truth, I am very disappointed.

The first part of this chapter was written from the perspective of being the Chairman of AMEC, a very large engineering and construction company. Having subsequently retired from my old company, I am now privileged to enjoy a somewhat less frenetic lifestyle, encompassing many different activities. Interestingly, in almost all of them, my role is that of client: whether as Chairman of English Partnerships, as a Director of CapitaLand in Singapore, or as Chairman of Manchester Millennium or the newly emerging regeneration companies here in the UK.

Perhaps, therefore, my point of view has shifted in that the point from which I view has moved quite fundamentally. My conclusions, however, are no different than they were five years ago. After the common-sense approach of Michael Latham, widely accepted in the industry as a whole, we then had John Egan's 'down to earth' appreciation of what we might do to improve the industry's performance.

We have seen much growth in partnering and alliancing, and good progress on PFI initiatives across a much wider range of investment areas. There has been some improvement in that time, but certainly not enough.

The construction industry is still nowhere near as efficient as it could be. I had hoped that the changes and improvements would have accelerated much faster but they have not. The number of major contractors in the UK has continued to reduce, partly by negotiation but also partly by attrition. The same can be said for designers and suppliers.

We do have world-class architects and engineering designers; we do have very able construction and manufacturing industries; but we are still nowhere near as good as we should be in bringing those high-quality components together in a more efficient way.

Too often, we have an alliance or partnership between a client and a major contractor and its design team, but it does not pass downwards to the subcontractors and suppliers. Too often, there is a continuing wish to pass risk further and further down the chain to a level at which the risk cannot possibly be carried.

We still do not spend anything like as much time at the concept and early design stage. The rush to begin is still too common, as distinct from a determination to complete within the prescribed time, within the prescribed budget and to the required quality.

Integration of the design process has improved substantially, but not sufficiently. The process is controlled very much by the weakest link in the chain. There remains insufficient commitment from clients to require full integration between their own and the various different design teams. Full integration requires high-quality leadership and a fundamental understanding by all the parties of each other's responsibilities.

Five years ago, the profitability of contractors from their normal contracting businesses was under great pressure. Today it appears they are under even greater pressure. The major public statements by all of them emphasize their commitment to reducing or indeed removing their involvement in traditional contracting in favour of exclusively negotiated or 'partnered' projects. But nowadays profit pressure on the design teams has become even more acute. Most certainly in all of these companies the financial rewards are entirely inconsistent with the quality of the intellectual rigor which is applied to all the projects with which they are concerned.

## Redefining the design team

In Chapter 3, Peter Rogers argues for the key role of the client in the design team as the innovator and the major risk taker. He is part of a very forward-thinking and efficient company. Would that all clients operated in the way that that company does.

The purpose of this, though, is not really to focus on individual buildings or even groups of buildings. It is concerned with the design team within the built environment as a whole. Here I do believe that major changes have begun and will inevitably change so much in the future. We are now seeing some remarkable new initiatives emerging, bringing together comprehensively for the first time the best talents of planners, architects and engineers, looking at land use, planning and transport planning, and at the social, educational, security and broader environmental issues surrounding our major urban centres.

I refer, of course, to the creation of the new urban regeneration companies. In June 1996, I was privileged to be asked by the then Government and the Leader of the City Council in Manchester to chair the special task force set up to rebuild the city centre of Manchester after the IRA bomb attack. As a result of an international competition, a master-planning team was selected, transport planners appointed and a wide range of architects and consulting engineers retained to make a fundamental difference to a city centre in a way no one had ever attempted since the end of the Second World War. It was a remarkable success, with some £100m of public sector investment resulting in further investments of more than £1 billion. It made a real difference. It was a most stimulating and satisfying experience, and even when the government of the day changed, within a year of the task force being established, it was a seamless transition, with central government and local government working together to complete this exciting renewal.

Since then we have had the report of Lord Rogers' Urban Task Force seeking a commitment to an urban renaissance, and it has begun. We now (in 2001) have three urban regeneration companies established: one in Liverpool chaired by Joe Dwyer, the former Chairman of Wimpey plc, one in Sheffield, chaired by Sir Peter Middleton, and one in East Manchester, which I am honoured to oversee. Three more are about to be launched and there are likely to be a total of some 16 over the next year or two.

In each case, they are a partnership between central government, the new regional development agencies and the relevant local authority, together, of course, with the private sector. Each of them requires proposals from integrated master-planning teams before selecting a preferred partner. That team, working closely with the Urban Regeneration Company itself, consults widely with industry and commerce and most of all with the local people. It is a proactive approach to planning, rather than what has become the normal reactive system.

More fundamentally, it is a holistic approach to an area, not an elemental, block by block mechanism. It works very well. It is 'joined-up thinking' between government departments at the centre, in the regions and with the local authority. They are, by definition, long term. They are not intended to be 'quick fixes'.

Given that I now view all of these matters from a somewhat different perspective than five years ago, I find myself in complete agreement with everything I thought then; and I would, of course, like to have seen things move ahead even more quickly, especially after Latham and Egan. What is missing is a much better understanding of the parts that each specialist must play in the creation of every project.

I am saddened to see the ignorance which often surfaces in understanding the risks and responsibilities of the parties to the creation of new projects. I am, however, greatly heartened by the opportunity to have in urban regeneration companies, integrated teams of experts, working in a holistic way to deliver long-term sustainable improvements in the built environment of so many areas of the UK. The processes are new, exciting and in my view inspirational. Happily, there is also much to do.

As always, the most important part is the design and, in particular, the concept which precedes it. Redefining the design team is therefore even more important now than it was five years ago, and I suspect that five years from now the message will be the same, although the balance will, as ever, continue to be different.

## CHAPTER THREE

# The client and the design team

*Peter Rogers*

### The client

Stanhope is a company that builds, and whether the client is internal or external, the company exists to serve this client. I emphasize this because I feel that the client is key for any project, not because the client is paying the money, but because it is the client's vision that is creating the project in the first instance. The vision may be naive and the client may be completely inexpert in building, but there is still a vision, and the client has a need. The client has to create something for its business – it may be a factory, an office building, a theatre or a motorway; it really doesn't matter – but that element of creation is the first step in the process, and it belongs to the client.

There has been a great deal of talk about putting the client in the driving seat, but I am not strictly an advocate of that. I believe the important point for a client is not who controls the process day to day; that is a secondary consideration. The client has to have the creative vision and has to translate that into a brief. This may be difficult because of the large number of people willing to tell the client how to write this brief. This is not an argument against the early involvement of designers, but it is an argument for the clear-minded formulation of the client's fundamental needs *before* design begins.

Without this focus, projects may become engineering-led or architecture-led, according to the bias of who is at the table. I have been doing review work for the Arts Council and the National Heritage Memorial Fund, and it is amazing how many briefs have been generated out of a design concept rather than from a clear understanding of the client's business objectives. The brief has to be researched and developed on the basis of these objectives, and this has to come from the client – he or she is the only person who really knows what the 'product' has to achieve. If help from professionals is required this should come from an individual who has no vested interest in the outcome and who may not be involved at all after the brief has taken shape.

The client must ensure that the design team has very clear responsibilities and clear guidance through this brief. It is only then that the energy of the design team can be directed towards meeting the client's need. At Stanhope we have looked at the way in which projects have been put together in the past and the problems that have been created by the lack of clear definition. In this history lie cautionary tales. We did some research several years ago looking at various projects that have been built by different developers and clients (one-off clients in particular). One example was a £90 million, 24-month project which turned into a £150 million, 42-month project. More important in a way, because that was an example of a fairly straightforward failure, was the need of another client to leave a building four years after moving in, not because the client had gone bankrupt, but because the business circumstances had changed. Suddenly, the large, purpose-built office building no longer fitted the client's business requirements. What is more, because the building had been overly bespoke, it was by no means easy to sell on. The property had thus become a loss, rather than an investment, the minute the client had moved in. In both of these examples a lack of definition of the client's true needs and a lack of control over the direction that design took in meeting these needs led to expensive failures.

The client also carries the risk in the project. The client has to identify it; it cannot be transferred. There is no such thing as risk transfer; it will always come back to the client, one way or another. Foundations are a good example. We could on the basis of a limited geotechnical survey offer a contract which passes the risk to the contractor. We then have two ways of spending too much money. Either the contractor covers the risk of the unknowns through a high tender or, if the contractor puts in a low bid, we end up paying through delays or claims when the contractor tries to cover its losses. With a large portfolio like ours we could become our own insurers and absorb the risk, but this is not our method. Risk is usually the product of unknowns, and the better way to eliminate unknowns is to do some research – in this case, by carrying out a more detailed survey. There is no such thing as an unquantifiable risk.

Quantification and comparison help us to identify where events might be out of step with our expectations, and to this end we have created benchmarking documents. These illustrate an unusual mixture of buildings, which show little consistency in use but which enable elemental costs to be interrogated. Table 3.1 is an example of the way we analyze the cost of city office construction and use the information to create performance targets. In this example, projects A, B and C are our own buildings while 'Project X' is for a more basic building. These are all compared to our 'Cost Model', which is the price we are looking to achieve (or better of course). These particular examples do not show a great variety in the totals, but studies like this are useful in highlighting anomalies. There might be unexpectedly high costs for individual elements (which need explaining) or large differences in built cost for similar buildings. For example, we have built City of London buildings for £100 per square foot whilst others have been built for up to £200 per square foot. Has double the value really been created between one and the other? When you see that the £200 per square foot building has virtually no two cladding panels that are alike, it is fair to ask whether the additional cost reflects an improved investment or demonstrates a lack of control of design. We might also ask whether, for an estimated £800 per square foot, Portcullis House offers three times the value of an equivalent fitted out commercial building in central London.

## Management of the project and project management

So how does one solve some of these problems (of cost, suitability and fitness)? I come back to the client – you must have a client, and it does not matter how expert the client is: that client has to be single-minded, must be a patron, and must not be a substitute or a committee – neither work. He or she has to be the individual in the organization who has the authority, the vision, and the financial muscle to make that project happen. The client also needs a supporting team.

Ian Ritchie speaks about the absence of barriers in collaboration – the team has to be brought together without barriers. Quite often this is done through one of my favourite hobby horses – the 'traditional' project manager. I have a genuine concern about project managers, even though I carry out the role myself on some of our projects.

Most project managers have come from a very narrow experience base and have little creative vision. This is not a specific criticism of them as individuals, but of the way our industry trains people and brings up very mechanistic managers. A project manager can place him- or herself as a barrier between the client and the design team, and once that happens you are doomed. I do not believe that the project is then likely to develop into a successful aesthetic project (it may be possible to have the more mechanical projects resolved this way because they are solely driven by technical issues). If you want creative, visionary projects there has to be a very close relationship between the client and the professional team. That relationship has to be managed of course; I'm not advocating that we end up with anarchy, but it is very important that the project manager does not prevent this creativity. Project managers should not make decisions on behalf of clients, but

*Table 3.1 Typical comparison of cost model with historical data and project cost projections.*

| Cost element | Project A (Out turn) 220,000 sq ft 1998 | | | Project B (Cost plan) 460,000 sq ft 1998 | | | Project C (Cost plan) 440,000 sq ft 1999 | | | Project X (Cost plan) 300,000 sq ft 1998 | | | Cost model (£85 sq ft) 250,000 sq ft 1998 | | |
|---|---|---|---|---|---|---|---|---|---|---|---|---|---|---|---|
| | **Cost** | **£/sq ft** | % | **Cost** | **£/sq ft** | % | **Cost** | **£/sq ft** | % | **Cost** | **£/sq ft** | % | **Cost** | **£/sq ft** | % |
| Substructure | 1,497,647 | 6.86 | 7.27 | 7,113,582 | 15.49 | 14.69 | 3,165,340 | 7.20 | 7.67 | 1,033,232 | 3.43 | 4.15 | 886,950 | 3.54 | 4.17 |
| **Substructure subtotal** | 1,497,647 | 6.86 | 7.27 | 7,113,582 | 15.49 | 14.69 | 3,165,340 | 6.86 | 7.67 | 1,033,232 | 3.43 | 4.15 | 886,950 | 3.54 | 4.17 |
| Superstructure | 3,341,074 | 15.31 | 16.23 | 6,642,831 | 14.47 | 13.72 | 6,806,759 | 15.49 | 16.50 | 4,782,448 | 15.88 | 19.20 | 4,067,440 | 16.25 | 19.12 |
| Fire protection | 545,375 | 2.50 | 2.65 | 1,227,080 | 2.67 | 2.53 | 415,010 | 0.94 | 1.01 | 281,340 | 0.93 | 1.13 | 498,200 | 1.99 | 2.34 |
| **Superstructure subtotal** | 3,886,449 | 17.80 | 18.88 | 7,869,911 | 17.14 | 16.25 | 7,221,769 | 16.43 | 17.50 | 5,063,788 | 16.82 | 20.33 | 4,565,640 | 18.24 | 21.46 |
| Cladding external | 3,996,782 | 18.31 | 19.41 | 8,773,855 | 19.11 | 18.12 | 6,915,309 | 15.74 | 16.76 | 4,567,310 | 15.17 | 18.33 | 3,067,500 | 12.26 | 14.42 |
| Atrium roof | 0 | 0.00 | 0.00 | 178,500 | 0.39 | 0.37 | 294,052 | 0.67 | 0.71 | 0 | 0.00 | 0.00 | 110,000 | 0.44 | 0.52 |
| Atrium cladding (int) | 223,000 | 1.02 | 1.08 | 1,093,537 | 2.38 | 2.26 | 683,228 | 1.55 | 1.66 | 0 | 0.00 | 0.00 | 0 | 0.00 | 0.00 |
| Roofing | 560,916 | 2.57 | 2.72 | 305,785 | 0.67 | 0.63 | 648,862 | 1.48 | 1.57 | 266,804 | 0.89 | 1.07 | 983,500 | 3.93 | 4.62 |
| **Enclosure subtotal** | 4,780,698 | 21.90 | 23.22 | 10,351,677 | 22.54 | 21.38 | 8,541,451 | 19.44 | 20.70 | 4,834,114 | 16.05 | 19.40 | 4,161,000 | 16.62 | 19.56 |
| Blockwork | 461,044 | 2.11 | 2.24 | 354,670 | 0.77 | 0.73 | 409,779 | 0.93 | 0.99 | 320,168 | 1.06 | 1.29 | 421,414 | 1.68 | 1.98 |
| Drylining | 725,817 | 3.32 | 3.53 | 898,204 | 1.96 | 1.85 | 1,090,708 | 2.48 | 2.64 | 987,063 | 3.28 | 3.96 | 625,016 | 2.50 | 2.94 |
| Finishes | 1,290,450 | 5.91 | 6.27 | 2,687,813 | 5.85 | 5.55 | 1,200,263 | 2.73 | 2.91 | 1,135,129 | 3.77 | 4.56 | 1,278,970 | 5.11 | 6.01 |
| Joinery | 842,340 | 3.86 | 4.09 | 762,187 | 1.66 | 1.57 | 1,036,693 | 2.36 | 2.51 | 687,628 | 2.28 | 2.76 | 559,048 | 2.23 | 2.63 |
| **Internals subtotal** | 3,319,651 | 15.21 | 16.12 | 4,702,874 | 10.24 | 9.71 | 3,737,443 | 8.51 | 9.06 | 3,129,988 | 10.40 | 12.56 | 2,884,448 | 11.52 | 13.56 |
| Water systems | 1,092,043 | 5.00 | 5.30 | 3,693,390 | 8.04 | 7.63 | 2,852,590 | 6.49 | 6.91 | 1,382,599 | 4.59 | 5.55 | 1,147,000 | 4.58 | 5.39 |
| Air systems | 457,918 | 2.10 | 2.22 | 0 | 0.00 | 0.00 | 1,479,604 | 3.37 | 3.59 | 1,562,060 | 5.19 | 6.27 | 1,023,336 | 4.09 | 4.81 |
| Electrical | 1,036,831 | 4.75 | 5.04 | 2,030,179 | 4.42 | 4.19 | 2,136,669 | 4.86 | 5.18 | 2,043,587 | 6.79 | 8.20 | 1,162,700 | 4.65 | 5.46 |
| Fire engineering | 244,908 | 1.12 | 1.19 | 430,909 | 0.94 | 0.89 | 293,629 | 0.67 | 0.71 | 231,316 | 0.77 | 0.93 | 186,032 | 0.74 | 0.87 |
| Controls | 257,599 | 1.18 | 1.25 | 531,178 | 1.16 | 1.10 | 388,317 | 0.88 | 0.94 | 100,946 | 0.34 | 0.41 | 230,000 | 0.92 | 1.08 |
| Lifts | 934,938 | 4.28 | 4.54 | 2,329,578 | 5.07 | 4.81 | 1,656,543 | 3.77 | 4.01 | 1,115,000 | 3.70 | 4.48 | 1,000,000 | 4.00 | 4.70 |
| Specials | 160,959 | 0.74 | 0.78 | 613,928 | 1.34 | 1.27 | 343,186 | 0.78 | 0.83 | 170,000 | 0.56 | 0.68 | 400,000 | 1.60 | 1.88 |
| BWIC | 41,914 | 0.19 | 0.20 | 500,000 | 1.09 | 1.03 | 810,324 | 1.84 | 1.96 | 304,522 | 1.01 | 1.22 | 232,540 | 0.93 | 1.09 |
| **Services subtotal** | 4,227,110 | 19.36 | 20.53 | 10,129,162 | 22.06 | 20.92 | 9,960,862 | 22.67 | 24.14 | 6,910,030 | 22.95 | 27.74 | 5,381,608 | 21.50 | 25.29 |
| External works | 71,800 | 0.33 | 0.35 | 144,398 | 0.31 | 0.30 | 250,835 | 0.57 | 0.61 | 717,200 | 2.38 | 2.88 | 50,000 | 0.20 | 0.23 |
| Utilities | 165,000 | 0.76 | 0.80 | 218,753 | 0.48 | 0.45 | 341,439 | 0.78 | 0.83 | 200,000 | 0.66 | 0.80 | 200,000 | 0.80 | 0.94 |
| **External elements subtotal** | 236,800 | 1.08 | 1.15 | 363,151 | 0.79 | 0.75 | 592,274 | 1.35 | 1.44 | 917,200 | 3.05 | 3.68 | 250,000 | 1.00 | 1.17 |
| Staff costs | 722,000 | 3.31 | 3.51 | 1,803,000 | 3.93 | 3.72 | 3,680,836 | 8.38 | 8.92 | 620,000 | 2.06 | 2.49 | 1,126,500 | 4.50 | 5.29 |
| Preliminaries | 1,516,101 | 6.95 | 7.36 | 2,897,000 | 6.31 | 5.98 | 1,843,476 | 4.20 | 4.47 | 800,000 | 2.66 | 3.21 | 659,298 | 2.63 | 3.10 |
| Fee | 380,200 | 1.74 | 1.85 | 876,422 | 1.91 | 1.81 | 867,692 | 1.97 | 2.10 | 453,414 | 1.51 | 1.82 | 360,000 | 1.44 | 1.69 |
| **Organisation subtotal** | 2,618,301 | 11.99 | 12.72 | 5,576,422 | 12.14 | 11.52 | 6,392,004 | 14.55 | 15.49 | 1,873,414 | 6.22 | 7.52 | 2,145,798 | 8.57 | 10.09 |
| **Contingency** | 20,807 | 0.10 | 0.10 | 2,320,043 | 5.05 | 4.79 | 1,650,000 | 3.75 | 4.00 | 1,150,000 | 3.82 | 4.62 | 1,001,372 | 4.00 | 4.71 |
| **Total shell and core** | 20,587,463 | 94.31 | 100.00 | 48,426,822 | 105.45 | 100.00 | 41,261,143 | 93.90 | 100.00 | 24,911,766 | 82.74 | 100.00 | 21,276,816 | 85.00 | 100.00 |

with them. They should not filter the communication between clients and design teams, but should encourage and manage it, acting as a catalyst and enabler for the project.

The other issue with most project managers is that they have never really been involved in taking risks, and that makes a big difference. This is probably the biggest change that I noticed in moving from construction to development. I worked for a contractor for much of my early career and I thought I took a lot of risks, but although every project a contractor builds contains risks, they do not compare with those carried by clients. The risks they take are enormous, because they take the total risk of the project. Not only is the specific construction at risk, but it is also likely to affect the future of the client's company for many years to come. For project managers (or contractors or designers), once they have finished a project they move on. They have liability issues and such like, but they are really secondary.

Management is needed, however, as, fairly obviously, the client needs to be a very senior person and will have far wider responsibilities than the construction project. The client must therefore have a clear management structure beneath him or her and, for the project to be successful the involvement of other people in the organization is necessary. The client has to engender a sense of shared ownership in the whole team. Part of the brief-taking role is obtaining information from other people in the organization, and for this information to be reliable and timely, they must be in sympathy with the projects objectives. How many projects go wrong because somebody has a vision at the top and the people beneath destroy it because they either do not believe in what is being created or the ambition has not been communicated to them? The CEO of an organization may not be able to be involved in the day-to-day administration of the project (he or she may attend only the key design team meetings) so it is therefore essential that someone within the organization takes on that part of the client's role. This is not an independent project manager's role.

If such a person cannot be found within the client's organization, the next option is to head-hunt someone who understands the business, someone to help the client develop the brief, help manage the process and maintain the presence of the client's core vision at the centre of the project.

## Building the team

Before design begins the client must have put in place the decision-making framework, should have established responsibilities within his or her own organization, and should have prepared a well-researched brief. The ground is then prepared to allow the immense creative skills of the professionals to be brought to bear on the project while making sure that the client and the client's ideals remain their focus.

Once design begins the different professionals and their vested interests have to be managed. Achieving a balance between these different motivations and skills is critical. It is not that any of the individual members are not capable; it is a question of harnessing their intellects to the benefit of the project by establishing an effective team structure. I am not talking here about contracts, because I really believe they are secondary; you only need a contract when things go wrong. This is really about attitudes and philosophy. The contract sets the rules of the game, but the important thing is to position the professionals and contractors as equals – professionals and contractors selected on the basis of skill and not money. Fee competition really does not work, and somehow we have to broadcast that message more widely than has been done in the past. Skilled designers, for example, provide the intellectual leverage that creates value, and it is wrong to argue over a percentage point here and there.

Design also needs time. Architects, for example, are required to do two things: one is simply to produce contract drawings, and the other is to use their skill and intelligence in solving problems. This needs time – time for research and for developing a vision. Cheapening the architect's fees means that they will spend proportionately more time doing the process work of churning out drawings and will spend less time thinking about what is being drawn, but it is the latter that offers the client most value. Designers need to make this case. They need to be able to negotiate from a basis of their knowledge of what design costs and how long it takes.

We must not allow the design to progress independently from the input of other members of the team, including the construction aspects. Once the team becomes unbalanced, you start going down the slippery road to creating unsolved problems. Everybody involved needs to understand what they have to produce and the framework in which they are going to produce it. We expect our designers to deal with our preferred suppliers who we engage early to obtain their input, although it would be fair to say that some designers and some suppliers are better at working together than others. This is not an entirely new idea, of course, as before the advent of contract management consultants tended to have a preferred stable of tenderers with whom, over time, they would develop a rapport. This may have been beneficial, but the relationships in our view became rather too cosy, with contractors doing work for consultants who then had implied obligations towards them. We are again talking about establishing closer links, but these need to be a little sharper than they have been traditionally.

We have always used construction management; it is a contractual form, if you want, but it is actually more about a philosophy. It gives the client direct access to all the team, not just the professionals but also the contractors. Yes, it involves a risk, but that really does depend on the competence of the professional team and the construction manager as to how much risk is transferred to the client. If they are all supportive of that process they can take a great deal of the workload and risk from the client. Part of it depends on having fully coordinated appointments with clear duties – something current professional appointments and current contracts do not do well.

A good example of this concerns a large job of more than £100 million with which we were involved. We found out, unfortunately too late, that the client had appointed the architects on a different form of contract from the engineers and contractors, and there was no responsibility for the design information to be procured to suit the construction programme. That sort of situation only becomes critical when things go wrong, but obviously a project that tends to have those problems is more likely to go wrong in the first place.

One must also look at payment terms. Much of the discussion going on in the market-place at the moment, following the Latham and Egan reports, is about how money is transferred. The construction industry is very bad at paying and tends always to try to do deals to cut down the price, but of course the money paid at the beginning of the job sets the tone for the remainder. Payment should be used as an encouragement. It should be used to motivate. It should not be used as a punishment. We have, for example, considered offering bonus payments to designers or suppliers who produce savings, but it has to be admitted that it is very difficult to administer this – the reason being that in a successful project, with everybody contributing well, it is very difficult to distinguish who exactly had the good idea, and the last thing we want to create is a divisive, point-scoring atmosphere.

One of my concerns with respect to partnering (and it is very similar to my concern about in-house experts) is that relationships become stale. There is a very narrow balance between a partnering arrangement (that goes on for many years) and an in-house company team that have also been there for many years. We have not found an easy way round it. We would like to work with the same groups of people (we are quite clear that there is a learning curve to get over) and we do tend to use a pool of designers that we work with repeatedly. But one has to invigorate these relationships, otherwise the inventive spark can be lost. One technique that we use is to hold project reviews which include people drawn from other parts of the companies working on our project as well as people who are themselves working on the project. In this way we widen the spread of influences and pick up ideas that may have appeared elsewhere.

The advantage that we see in these stable relationships is the reason that I have found EC regulations problematic in the past, because if one follows the rules to the ultimate end, you would have to employ teams that were selected on a very random basis, and I think that is greatly increasing the risk. There is a need for teams to understand each other, and there is great value in working with teams that have an empathy for each other. We try not to change more than one member of a team on any one project, so that the other members of that team can help support the new member. It is very important to establish a stable base that allows change to be managed safely.

## Making progress

Just a brief point on quality, which I see not in terms of ISO 9000 or BS 5750, which are really about making the same product repetitively. They do not address whether that 'thing' is of any value, whether it works or whether it is of any real use to that project. Our industry is very much about innovation, and one of the dangers of standards is that they tend to close down our ability to innovate. The problem that we have in our industry is to balance innovation with the needs of cost and programme. One has to look at ways of using standard products and avoiding prototyping, but there has to be innovation, because without innovation our industry would slowly be stifled. Most projects have highly repetitive elements and one is left dealing with only the remaining 10% in which to apply innovative ideas. One way we have found of trying to engender good, new ideas is by benchmarking – identifying ideas, particularly in other industries, and seeing how they can be applied to construction.

There is a danger, though, that innovation can be seen as a desirable end in itself rather than a means to an end. Design that pushes too hard to be novel runs the risk of leaving the contractors and suppliers behind. Innovation can ask more of the industry than it can comfortably deliver.

Because of such risks it is best only to innovate if you have time and money to recover from the ensuing problems. Our preferred way comes out of our long-term relationships, whereby we can agree with a contractor that they might try something out on a project of ours that they have come across on another job. We might even operate a profit share so that if there are cost advantages, both the contractor and ourselves can benefit. If, on the other hand, the idea is not successful we can cover the cost so that the contractor does not suffer. This of course means we operate an 'open book' type of negotiation and also means that we need to work with people whom we have learned we can trust.

Innovation is not limited to design ideas or new products. We also consider innovation in project processes and communication. I am a great believer in getting the design team to 'walk me through' the building while it is still at the design stage. How does this space work? How does one element join to another? I am not, however, an advocate of 3D computer modelling just for visualization, because I am not sure how much I can trust what I am seeing. Is this a model or is it just a convincing picture? What we are doing in the way of computer modelling, which I find potentially very interesting, is the graphical modelling of the construction sequence. We are developing this on a current project in conjunction with Bovis Lend Lease, Architectural 3D and Unipart Coughlin. Together we are developing a logistics process for construction projects based on experiences gained in the motor industry. The logistics programme schedules all the components in their sequence in the construction and feeds this information to 3D modelling software. The combined package does not just model the building but models its assembly and gives us the chance to find glitches in the delivery or construction sequence (Figures 3.1–3.3).

## Changing attitudes

Construction needs these new ways of communicating and these new relationships, but while the Latham and Egan reports have done much to move the industry out of its complacency, neither has been fully successful nor totally accepted. Latham raised the right issues, but he raised too many to solve all in one go. Asking for so many changes at one time from so many people almost guaranteed that someone somewhere would object and delay progress. Egan, on the other hand, was more effective, but, in my opinion, did not have enough to say about design.

I also believe that neither Latham nor Egan stressed adequately one of the fundamentals, which is education. We do not provide adequate education at the moment; we train people to pass exams and get jobs. Once they get jobs then education starts, but by then many of our best students have passed the industry by.

I worry about architectural education. We have many of the best architects in the world, but there are too many being taught to be (or

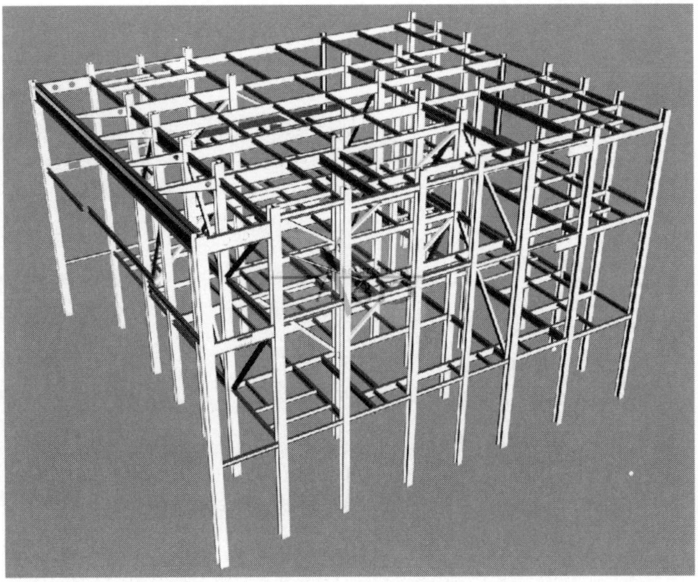

*Figure 3.1 While appearing to be a conventional model, this image was built on screen in the same sequence as the elements were to be assembled on site, offering a clear visualization of the construction process.*

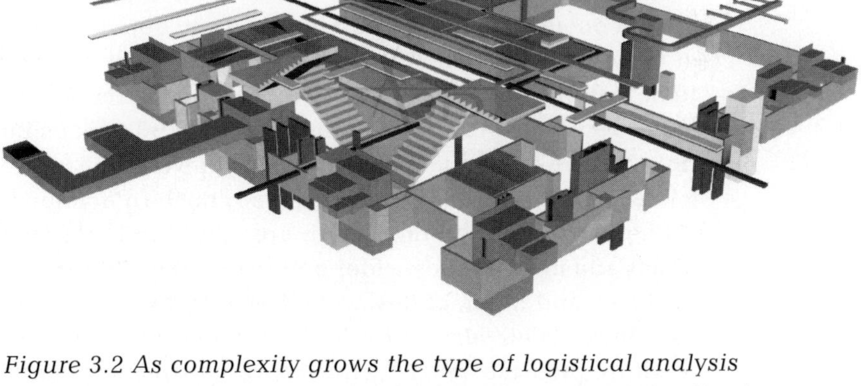

*Figure 3.2 As complexity grows the type of logistical analysis commonly used in the automobile industry can provide effective control of the flow of materials. Modelling the sequence of construction allows the coordination of components in spatial terms, but additionally enables the effective coordination of the contractor's work patterns.*

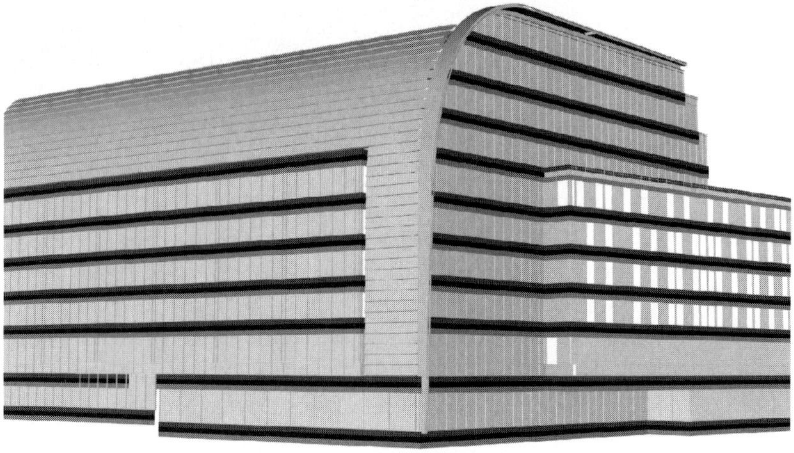

*Figure 3.3 Modelling the installation of this cladding in advance of construction revealed a potential hazard which was avoided by making changes to the programme of works. Construction was made safer without the disruptive changes to the flow of work that would have arisen had the problem been first noticed during construction.*

believe they could be) the next generation of top architects and there are not enough people being taught how to detail the technical aspects of buildings. We still spend a lot of money just preventing buildings from leaking. We get towards the end of construction and find far too many design errors that we then have to spend time and money putting right, such as flashing details that do not work or handrails that do not join at landings. This part of design may not be glamorous, but it is vital.

Engineers, on the other hand, are very tightly focused on technical issues and need a much wider understanding of what is going on in the world around them, of design and architecture and of how people use buildings. Their almost exclusively technical education from GCSE onwards puts too much emphasis on getting the mathematics right (important though that is) and leaves them out of touch with the ambitions of others on the team.

Now perhaps it would not matter too much if the strengths of one part of the design team made up for the weaknesses in others, but what we see too often is fragmentation and poor communication. I do not believe that this can be changed by systems and legislation. We have to change attitudes, and the only way we can change attitudes is by education. We have to broaden the education of the graduates that are coming out.

The Royal Academy of Engineering produced an interesting statistic which said that for an office building £1 represented the capital cost, £5 the running cost and £200 the people cost. The precise figures vary from job to job, but clearly to an organization occupying a building, its capital cost is insignificant, the running cost is much less important than the industry currently believes (accepting the need for environmental concern) and it is the people cost that truly matters. Clients are becoming much more aware of this and are becoming more involved in the philosophy of their buildings. Do they help people be productive? Can they reduce the turnover of staff? Can they help attract top-flight employees? These are issues that the design team must be encouraged to address.

We also have to change the vested interest of our institutions. I resigned from the Vice Presidency of the Institution of Civil Engineers because I came to the conclusion that it was an immense bureaucracy which probably did more harm than good to the industry. The whole qualification process was about qualifying people into the Institution of Civil Engineers, not about qualifying people into the process of providing a better built environment, be that a motorway, office building, theatre or city.

Professional institutions, in general, are still too narrowly focused, with too much emphasis on defending their professional patch. There may well be a bridging between the institutions at the higher level, but by then it is too late. It is not enough for senior members talk to each other. The communication has to be made to happen much lower down, and it should start with professional education. The institutions approve degrees and decide what knowledge and abilities make people qualified for membership. They wield great influence over the raw material that makes design teams. Some of this influence should be turned towards bringing the professions into closer contact, so that design teams can communicate and share a vision. I spoke on these matters at a meeting at the Royal Academy, which is one organization that actually manages to bridge very successfully across all professionals and that is a lesson we must learn.

## Conclusions

So how should we redefine the design team? We must identify the client – the patron – and I believe that it is everyone's duty to support the client. The client is not there just as the paymaster; he or she really is there as a focus, and how much support has to be given to the client depends on that person's experience. Anyone commencing a new project must ensure that the project has a client, and not accept second best. Many clients tend to turn round and say 'I'm too busy – I've got to run the business'. You are guaranteed at that point to have a problem. Without a unifying ambition, the project will be a battleground were the ambitions of others will fight for ascendancy.

We must support interdisciplinary design as strongly as we can, and courses such as IDBE at Cambridge are an essential part of getting our industry to communicate and to work together. We must get engineers to talk to architects: not just the few exceptional ones at the top, but every engineer and every architect and every cost consultant and every contractor must be able to talk to each other.

We must work to retain the creative vision of our best designers. Much good design is done by British designers overseas – why can we not get that level of patronage in this country? Part of it is oversensitivity to risk, because too many people in this country believe that good, exciting design will fail. This is a dreadfully negative attitude, as we are actually very successful at putting up great buildings and we are most successful when no member of the team becomes over-powerful. Successful projects are highly balanced projects, where everybody works as a team.

**CHAPTER FOUR**

# The design professions in transition – engineering and architecture in building for a small world

*Michael Dickson*

## Introduction

The traditional view of advances in architecture, engineering and the built environment is encapsulated in the quotation from *In Search of Man* by Andre Missenard (1957) and stated in *Climate and the Energy of Nations* by S. F. Markham (1942): 'Evolution of civilisation is closely linked to the potentialities of heating which are dependent not only upon the techniques of the apparatus but on fuel resources'.

Historically, development has been characterized by advancement in specific areas in a cyclical way – in 'structure' from masonry to frame buildings, from timber to steel to reinforced concrete into composite construction. In environment, windows led to the chimney, to Victorian ventilation systems, to gas lighting, to low-energy lighting; coal fires to central heating and air conditioning and displacement ventilation and chilled ceilings and back to historic 'passive' measures. Really it has been a history of sequential improvement, item by item.

Worldwide, we now need to vigorously pursue the concept of 'sustainable development'. As defined by the Brundtland World Commission in 1987 on Environment and Development this is: 'Development which meets the needs of the present without compromising the ability of the future generations to meet their own needs'.

Since the Rio de Janeiro Conference in 1992, there has been greater pressure worldwide to ensure sustainable development. By this, I mean ensuring that the aesthetic, economic and social benefits of a development outweigh the disruption to the local environment, and that the effects of by-products – pollution and extraction caused by the development 'from cradle to grave' – does not in the long run reduce the 'natural capital of the planet'.

The 1997 Kyoto Summit committed to a 5.2% reduction in carbon dioxide emissions by the year 2010 for all developed nations, with the UK agreeing to cut 20% off carbon dioxide emissions. Energy use in buildings in the UK accounts for between 40 and 50% of these emissions. Of this, 20–30% could be saved through widespread application of cost-effective energy-efficient measures both in the regeneration of our existing stock and in the application of new design techniques to new projects and further reduction of impact by greater use of renewable non-polluting energy sources.

Truly sustainable architecture (small 'a') requires the resolution simultaneously of many complex issues. All interested parties in the entire design, construction, operation and decommissioning process of the built environment have to be concerned with the global environmental consequences of their actions. We need concern for our actions throughout the project/building life (and beyond). A much more holistic and interactive process is needed where all disciplines interact by bringing their specific rigorous knowledge from a particular discipline into the shared domain (Institution of Structural Engineers, 1999).

Such influences will affect both the stylistic outcome of new designs and the engineering devices adopted to ensure these efficiencies. The issues are complex and different for each building usage, so the language of sustainable architectural engineering needs to be expanded immediately. It applies to a diverse range of building types, from hospitals and high-rise towers to offices and universities and rural communities. It is a movement driven in part by the value being placed in the post Egan era on economic performance and in part by the need for less damaging environmental impact (Figure 4.1).

Put bluntly, this 'cradle to grave' approach requires sustainable architecture to be sufficiently engineered to embody minimum energy while allowing for adaptability of future use and flexibility of arrangement during the complete cycle of use, refurbishment, reuse and recycling. To this end, building services installations have to be conceived to consume the minimum amount of high-grade non-replaceable energy consistent with comfort. There needs to be a distinction between stored energy and high-grade 'power' requirements. We need to study solutions from the past and join them to modern computing power and technology to evolve a new architectural and engineering language that will benefit the occupier and reduce the environmental score (kilograms of carbon dioxide per square metre per year) and so on for following generations. This will produce in turn its own architectural engineering aesthetic.

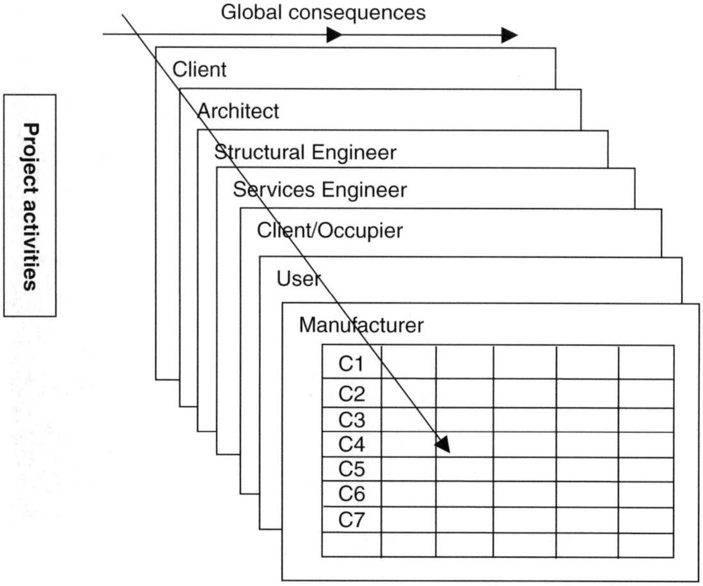


*Figure 4.1 Holistic working towards minimum global consequences.*

## Engineering architecture for a small world[1]

Architecture has to be purposeful, elegant, functional and affordable. We are, after all, talking about the conception, planning, modelling, calibration, construction, operation and use (over many years) of buildings where engineering has a part to play at every stage, in an architecture that has to be both 'delightful' and 'sustainable' (Figure 4.2).

As Le Corbusier rightly observed, good architecture needs a fusion of art and science so that the overlap of training between the two disciplines and the establishment of mutual trust is important. Architectural training has to capture society's art, spirit and philosophy, as well as excellent communication skills and a knowledge of building. To shape architectural developments, an engineer brings to the process observation, knowledge of scientific principles, methods of measurement and behaviour of materials. Engineers too must have the ability to communicate these principles in essential form to constructors. Of course, skills are interchangeable; so that all parties have to share the burden of achieving real delight, economy and functionality.

Post Cold War society is much more aware of the need to conserve the energy and mineral resources of the planet and to protect the natural

1 From a presentation at the 'Engineering architecture' conference, University of Glasgow, May 1999.

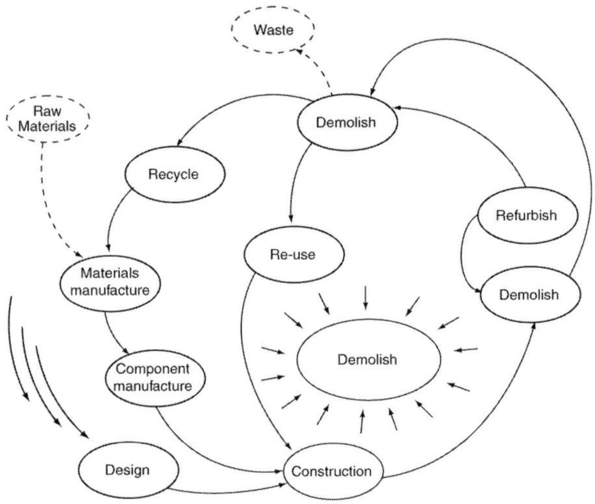


*Figure 4.2 Cradle to grave.*

resources that we now have left after the second millennium of the so-called civilized world. Better living conditions for more people makes the struggle for efficiency worthwhile – and yet economy and functionality must go hand in hand with architectural delight. In order to respond to these issues, architects, engineers and constructors need to radically rethink their approach to buildings and how they work together with clients – as proposed by Sir Michael Latham in *Constructing the Team* (1994).

In following this agenda, the creation and management of a healthy built environment has to be based on resource-efficient and ecological parameters. Specifically these issues are:

- appropriate location of development
- choice of built form
- choice of structural material
- minimizing the effects of construction and deconstruction by use of appropriate technology.

And these issues have to be understood in terms of their effects on the environment:

- amount of fossil fuel to purchase, process, transport and erect a building (embodied energy)
- use of non-renewable resources (finite materials, damage to environment, water)

- pollution (carbon dioxide, global warming, $SO_2$, $NO_X$, acid rain, CFC, ozone depletion)
- disruption to the local environment.

Architectural construction is once again becoming conscious of the need to ease the burden on the planet without loss of pleasure, value or comfort. This is being done by the process of improving economic performance as defined by Lower Life Cycle Costing and stimulated by considerations of less environmental consequence as indicated by Life Cycle Assessment.

## The Romantic versus the Classical approach – a useful tension

'We live in an era of continuous improvement and the next ten years must be better than the first. For this to be so we need to capture the imagination and leadership of a younger generation of construction professionals.' So says Graham Watts, Chief Executive, Construction Industry Council (Watts, 1998).

As an engineering consultant and a visiting professor of engineering design in a joint school of architecture and civil engineering, I often marvel at the range of historical knowledge and precedent my architectural colleagues possess. I am equally dismayed by my lack of awareness of historical developments in the use of engineering materials in construction (and by inference that of our engineering students). Yet what in Pirsig's *Zen and the Art of Motorcycle Maintenance* (1974) is contrasted as the difference between the 'Romantic' (based on precedent and immediate appearance) and 'Classical' approach (based on logic and underlying form) is, I believe, the very tension that can lead to a truly innovative modern architecture.

By this I mean an architecture whose planning and aesthetic come from the fusion of these different, often incompatible, threads of joint working (Rice, 1994). The outcome is an architecture whose aesthetic is often uncompromising and takes a while to gain acceptance. One such uncompromising new form was the Frei Otto design of timber grid shells at Mannheim (Figure 4.3). These enormous structures were initially described in *Die Zeitung* as 'two giant copulating whales' – and yet at the time the very innovation in engineering and architecture gave the design its own unusual exciting consistent aesthetic (Walker and Addis, 1997). The new engineering architecture may indeed come with its own aesthetic and impact.

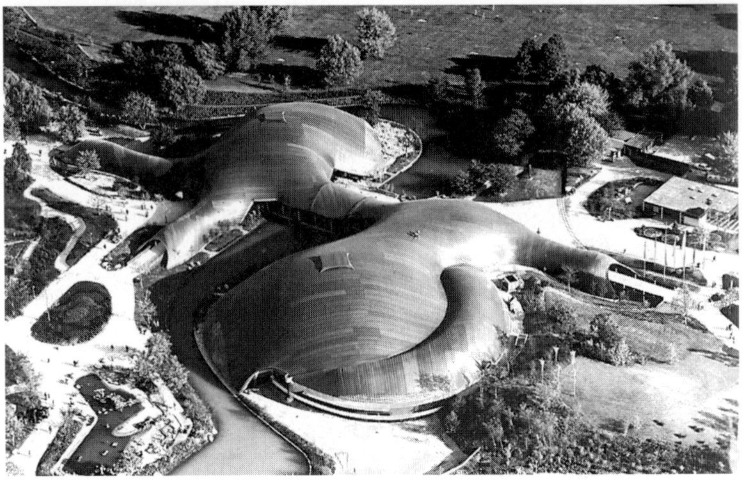

*Figure 4.3 Mannheim Bundesgartenshau – architects Muschler/Otto.*

## The changing process

There is still an architectural impact which comes from each building being a unique entity of design and place where each component has to be carefully selected and crafted for its particular place and duty (Groak, 1992).

### Worcester College

Designed before three-dimensional CAD modelling techniques were available, the Sainsbury building for Worcester College, Oxford (1978) relied on the conventional technology of the time. The design employed architectural plans and sections and was accompanied by isometric studies and the required craft construction to a very full set of working details (Figure 4.4).

The engineering of the building brought environmental layering of the interior spaces where the warmest space (the accommodation) was insulated from the outside by the cooler spaces of the corridors. Since the walls on one floor do not align with the walls at the lower floors, some fairly complex grillage analysis (at the time) of the intervening flat slabs in relation to construction sequence was required. Even though the building is essentially domestic in scale, careful consideration was required to assess the comparative stiffness of the wall supported on the flexible slab. Essentially, such buildings demanded of their designers and constructors extremely careful management to overcome the many

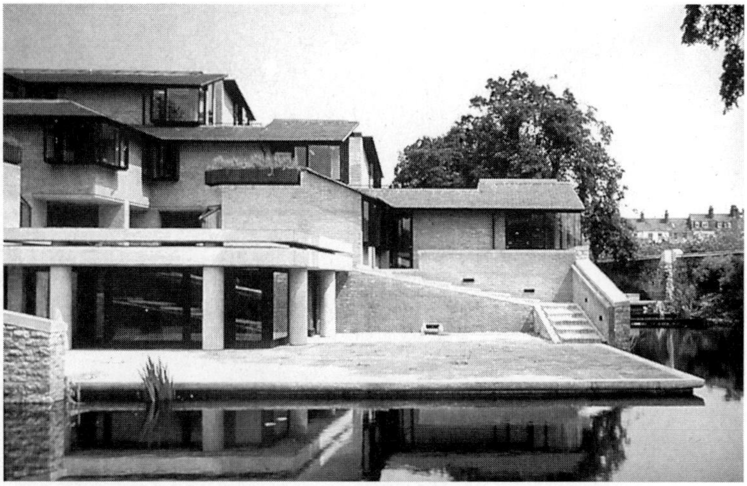

*Figure 4.4 Worcester College – architect Richard MacCormac.*

detail problems that are met on such projects. Craft building by small, highly specialized constructors is still a viable option even in the era of Rethinking Construction, Standardization and Factory Production.

**The Lowry Centre, Salford**

In contrast, The Lowry in Salford (Figure 4.5), with its 1650-seat Lyric Theatre, 450-seat adaptable theatre, two galleries and other ancillary space, benefited from the advances in three-dimensional modelling of structure and architecture and CFD environmental modelling of the interior volumes. The Lyric Theatre is wrapped on seven sides by a 28 metre high reinforced concrete wall (Figure 4.6), where the two flank walls and stage return are vertical but the front foyer walls slope out at a maximum of 17° to the vertical. Stairs punch through holes in the sloping wall, while cantilever balconies for the upper foyer and bars extend above the main foyer below. The 600 mm lower wall sections are highly stressed, but the upper section was reduced to 200 mm.

The balcony structure (Figure 4.7) is a steel frame which acts compositely with the concrete walls and is stabilized by flank wall bracing and the concrete flank walls tied by the floor diaphragms. After the concept was complete, extensive 3D structural analysis was provided to enable detail design and specification of each component as a 'finite element'. The steel frame balcony structure precast concrete seating units and ceiling in the Lyric Theatre form a plenum for air supply under the seats. Levels of precast units are set exactly in order to ensure a first-class

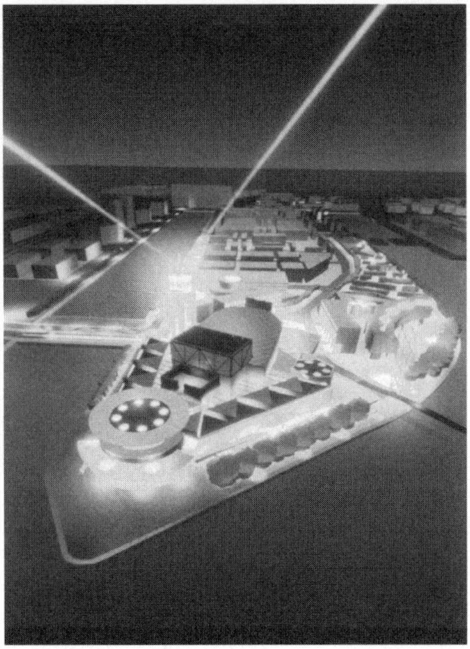

*Figure 4.5 External view of the Lowry Centre – architect Michael Wilford & Partners.*

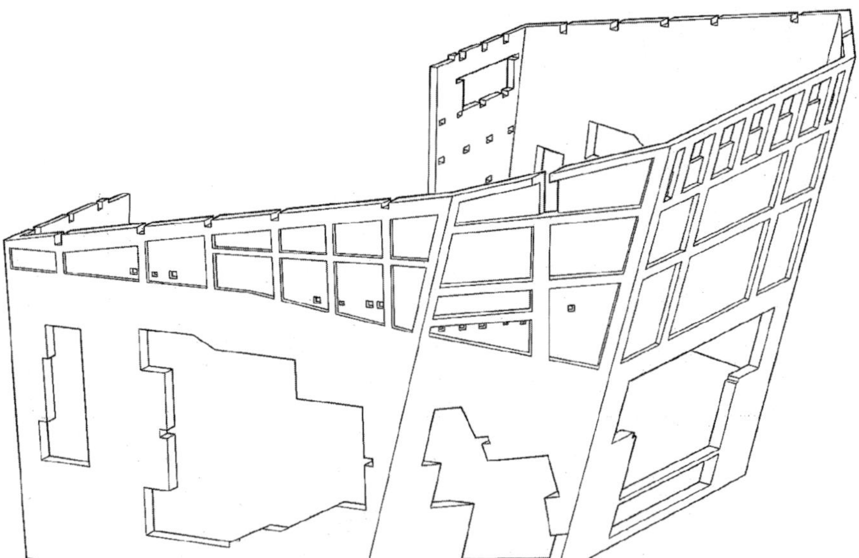

*Figure 4.6 Lyric Theatre – reinforced concrete wall.*

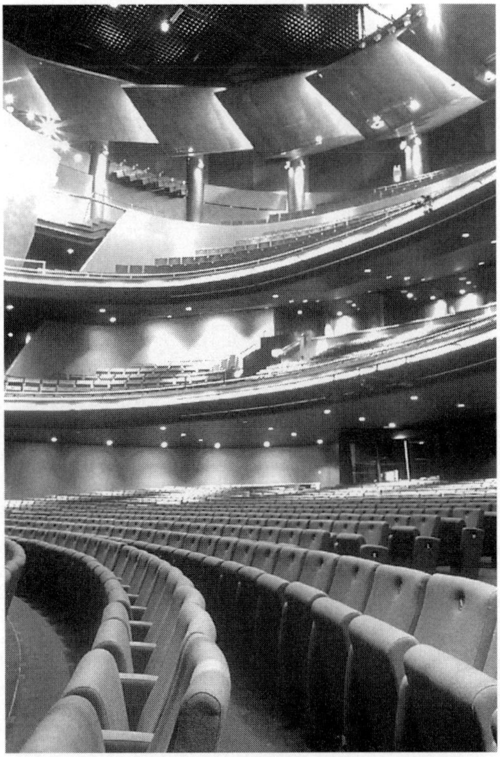

*Figure 4.7 Internal view of the main auditorium.*

sight line from every seat, each of which has been individually modelled in a three-dimensional design process by the theatre consultants.

Even after Egan's *Rethinking Construction* (DETR, 1998), most buildings have a unique identity with really no other equivalent. Even in the past five years the building design process could at best be an ill-defined iterative process where even though most drawings had been produced entirely using CAD, they were still only thought of as 2D plans. Coordination of the various disciplines was therefore a manual or mental operation because the design was still two-dimensional representations and not a true 3D representation. Gradually, as in the automotive industry, the construction industry is moving towards computer modelling and virtual prototyping of all components in the whole, enabling better standardization of the components.

### Centre for the Performing Arts, Bristol

The complex crystalline architecture enclosing the non-symmetrical, irregular, 2300 person concert hall for the harbourside in Bristol (Figure

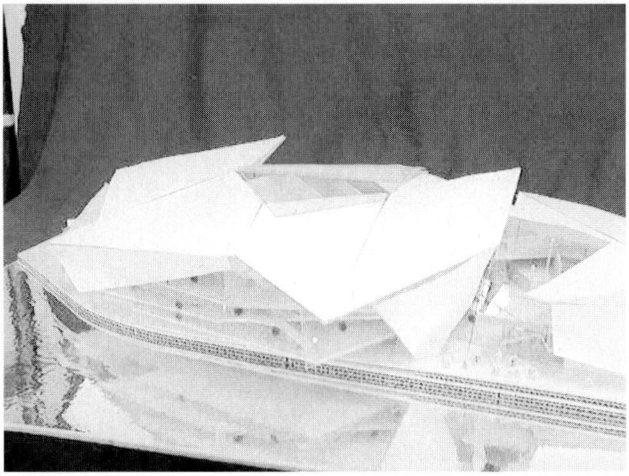

*Figure 4.8 Architect's model of the Bristol Harbourside Centre – architect Behnisch & Partners.*

4.8) designed by Behnisch & Partners relied substantially on CAD modelling. Definition of the solid geometry of each of the non-vertical, non-regular walls, floor plates, columns and so on was defined by individual finite elements within the global system geometry, and loads and structural actions could then be investigated both at the macro performance level and at the local detail design needs of each individual element. This virtual modelling was also accompanied by a careful, slow, painstaking construction of an equivalent physical structural model so that the engineers could compare the results from large-scale finite-element system analysis with more primitive hand-driven calculations, drawn from experience and from studying the physical model itself.

### School of Mathematical Sciences, Cambridge

A further stage in the process led to the modelling of the complex steel and reinforced concrete composite construction similar to that used for specifying the central conical roof at the Cambridge School of Mathematical Science.

This centre is on a green field site, bordered on three sides by residential property. Individual mathematical pavilion houses are arranged around the central core for meeting areas, laboratories and lecture theatres. The whole development is sunk within a landscape saucer thus allowing natural light and ventilation to the lecture theatres and laboratories to be located on the lowest sub-ground floor. This is an unusual if

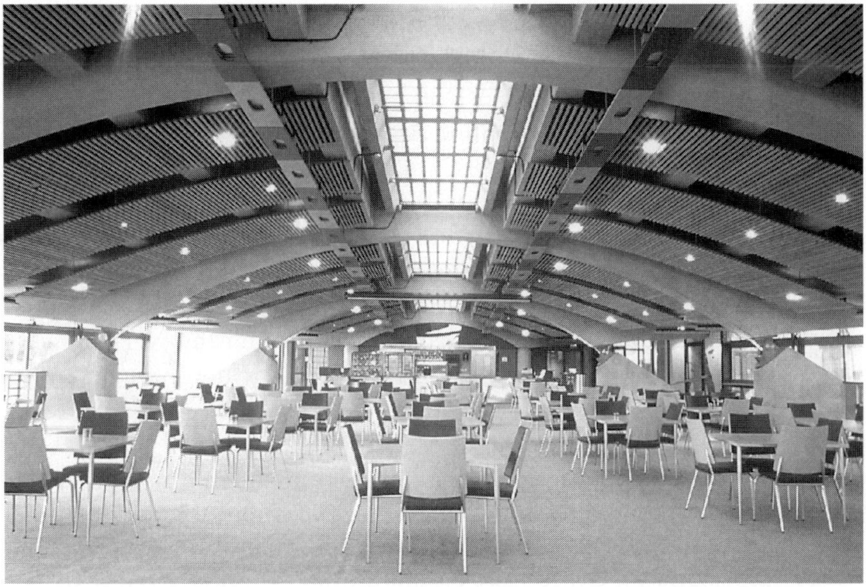

*Figure 4.9 Centre for Mathematical Sciences, Cambridge University: internal view showing cast bearing to arches – architect Edward Cullinan Architects.*

'sustainable concept' where the aim of this design is to replace the original open field site with an intense development of individual pavilions on a domestic scale in a landscape park. The central core has a relatively low eaves level and a fully turfed roof to minimize its impact on the environment.

The curved roof for the upper floor (Figure 4.9) spans a maximum 21 metres over the central meeting area and employs a series of arch structures at 8 metre centres over gradually increasing spans between identical buttresses. The shallow steel arched form resists the applied load from the turfed landscape roof and incorporates a roof light and ventilation slot and clerestory glazing. To set the construction geometry the primary arches and secondary beam structures are prefabricated in structural steel and were modelled as representative solid elements. Plate girder arches are arcs of circles each of different radius, subsequently encased in reinforced concrete. To express the springing point a 450/10 spheroidal graphite (SG) iron casting with 150 mm pin is used to transfer the 4000 kN ultimate thrusts. Because of the importance of the detail design of this pin, both as a load transmitter and as part of the visual response of the conical arch system, the detail form was studied by the engineer on a plasticine model which was accompanied by structural analysis of the solid form.

### Al Faisaliah Complex, Riyadh

The ability to study the interaction of structural form with likely patterns of the wind flow using complex CFD analysis is just one of the emerging techniques arising from the post-computer era which could bear on the costs and opportunities in engineering high-rise and other architecture.

In the desert climate of Riyadh, the three-dimensional relationship of the cladding and r.c. megastructure for the Al Faisaliah complex used elements of this three-dimensional modelling capability to define and coordinate particular zones for construction – for cladding, for servicing and structure and for glazing and shading.

One of the interesting developments that this project stimulated was detailed analysis of the wind flows and its loading on the large overhanging shading (Figure 4.10) to the effectively clear glass of the window. This uses computational fluid dynamics (CFD) and showed that the individual blades would be subject to considerably lower localized pressures than might have been postulated directly from the macro wind tunnel analysis. This resulted in a saving to the client in cost and improvement in the ease of detailing these important architectural elements. Indeed, three-dimensional prototyping was taken further in defining the megastructure and the coordinates for Wang Xiang International Plaza Tower, Shanghai (Figures 4.11 and 4.12).

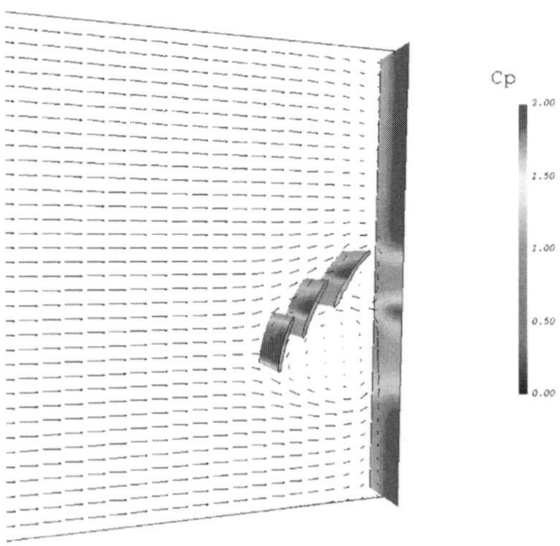


*Figure 4.10 CFD model of window shades, Al Faisaliah.*

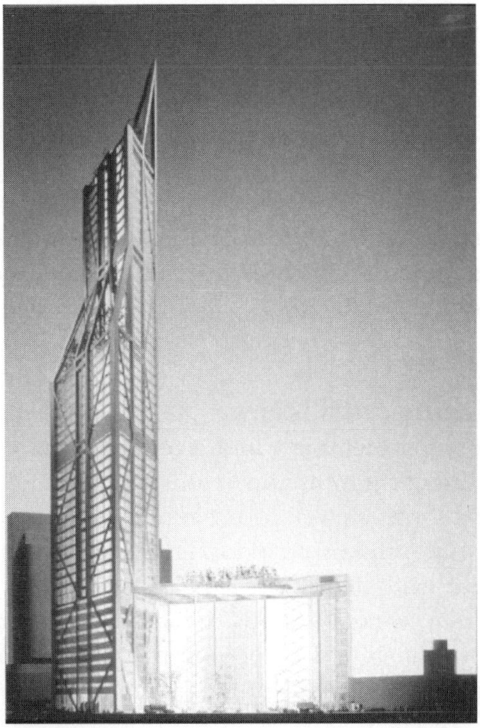

*Figure 4.11 External view of Wan Xiang Plaza – architect Ingenhoven Overdiek & Partners.*

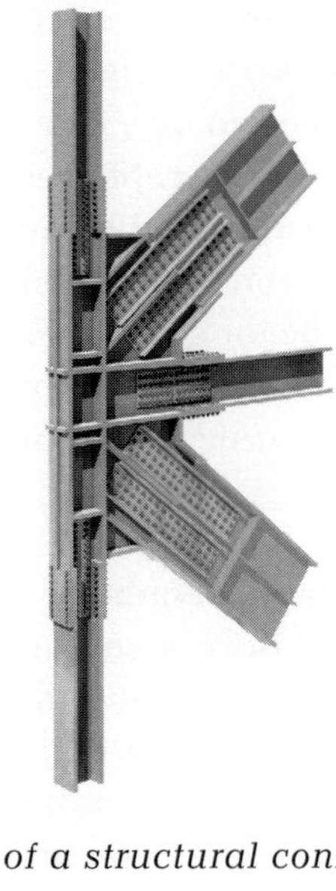

*Figure 4.12 3D prototype of a structural connection for Wan Xiang Plaza.*

Single building model or virtual prototype building (VPM) is the next logical step in CAD data production, which will allow coordination in the fourth dimension (time). This interoperability allows full exchange of data between practices and the attachment of non-graphical data to components – columns of concrete with specific grade and quantity of reinforcement which can be analyzed and costed accordingly. The single building model will allow automatic extraction of individual components for bills of materials, detail design, fabrication and assembly of details.

## Designing with form

Clearly structures which use their form to carry their loads principally by direct compression or tension offer the opportunity for exciting structural forms as well as minimum use of materials.

Microprocessing of form finding, analysis and patterning enables the rapid and detailed study of how surfaces act under loading. This increased comprehension has led to flatter structural surfaces which utilize the bi-axial elongation underload of prestressed fabrics to absorb the dynamic gust loadings upon them.

### Millennium Dome, Greenwich

The segmental double layer PTFE glass fibre fabric enclosing the Millennium Dome in Greenwich (Figure 4.13) demonstrates this use of patterned high-strength structural membranes. The 80 000 square metre stressed skin surface comprises 70 nearly flat segments of membrane supported by a post-tensioned cable net which is comprised of essentially straight cable sections between nodes. In turn, 24 radial cable lines are supported by an array of stay cables from the 12 lattice vierendeel masts on quadruped footings and anchored to the perimeter mast and ring beam structures. The resulting innovation of flat panel fabric and straight cable segments uses only 15 kg/m$^2$ of steel and cables and 1.5 kg/m$^2$ of fabric – less than the actual weight of the contained air.

### Stuttgart 21 Hauptbahnhof

Computer-aided design coupled with computer-initiated manufacture (CIM) is once again facilitating the design of massive vaulted forms where the benefits of structural mass acting as a thermal flywheel can be calculated from computational fluid dynamics (CFD). The design for the new Stuttgart 21 Hauptbahnhof (Figures 4.14 and 4.15) exploits this possibility. The 43 000 square metres of uninterrupted three-

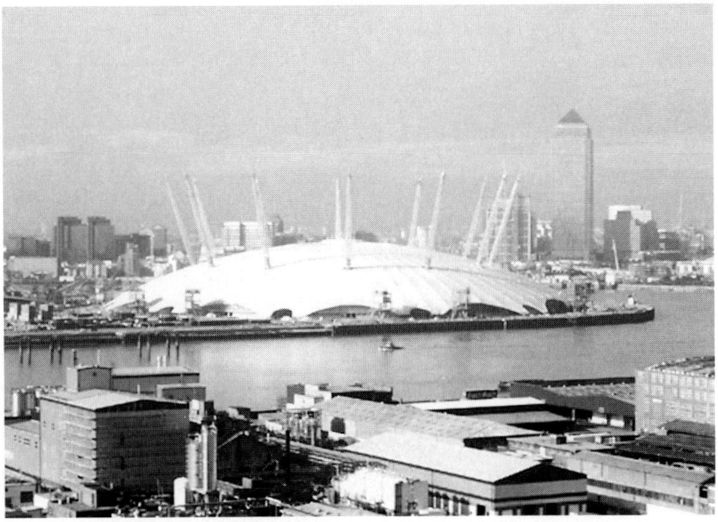

*Figure 4.13 External view of Millennium Dome, Greenwich – architect Richard Rogers Partnership.*

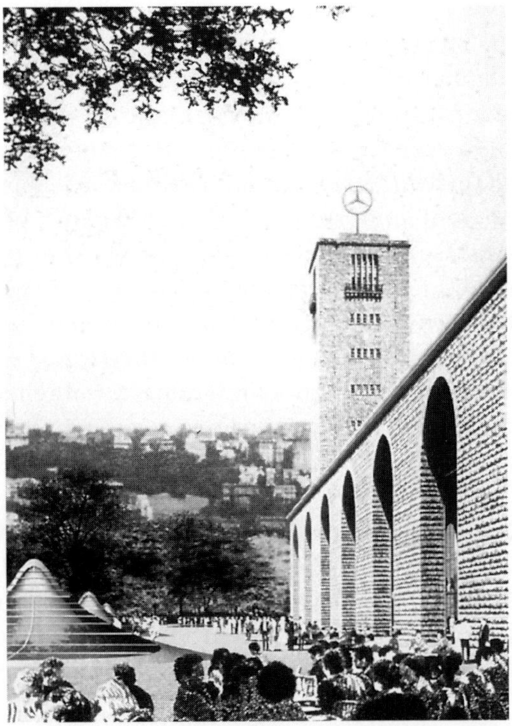

*Figure 4.14 Stuttgart 21: view of Bonatz Building standing on new station – architect Ingenhoven Overdiek & Partners with Frei Otto.*

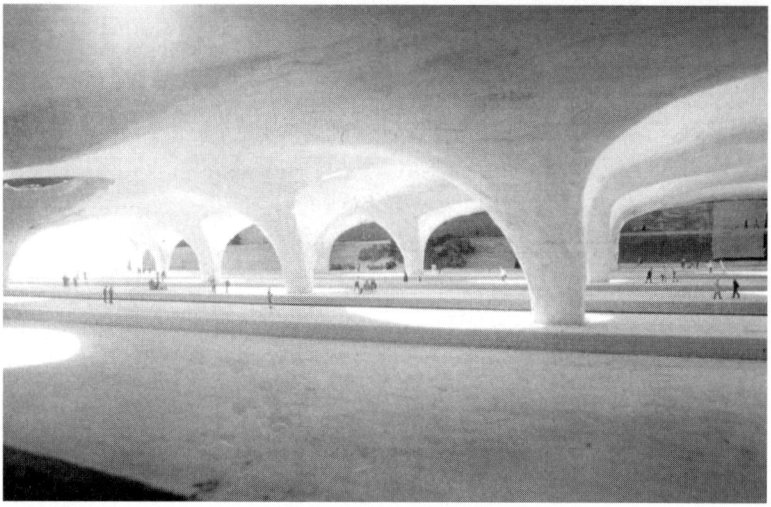

*Figure 4.15 Stuttgart 21 station form model – architect Frei Otto.*

dimensional flowing reinforced concrete shell forms is designed to produce a naturally lit environment where the temperature through the year remains stable in the range 15–18 °C due to the thermal mass of shells and the platforms and adjacent railway tunnels. The shells are 35 cm thick at the crown increasing to 65 cm at the light admitting eyes in the roof and supports the 25 $kN/m^2$ of loading from the historic Schlossgarten and new station plaza adjacent to the existing Bonatz building. The horizontal component of the shell thrust is resisted by a three-way network of post-tensioning cable net parallel to the top surface of the shells, rather as the strings of a tennis racket. The perimeter walls and their foundations have then only to resist overall stability forces from earthquakes.

## Exploiting materials science for architecture

Often quantification on cost parameters alone will lead to a radically different outcome in respect of sustainability than would occur if a qualitative assessment on environmental grounds from cradle to grave were also made. The Queen's Building is an example of this divergence.

### Queen's Building, Emmanuel College, Cambridge

The architect wished the new 200 seat concert hall for Emmanuel College (Figure 4.16) to be constructed from the same local stone as the

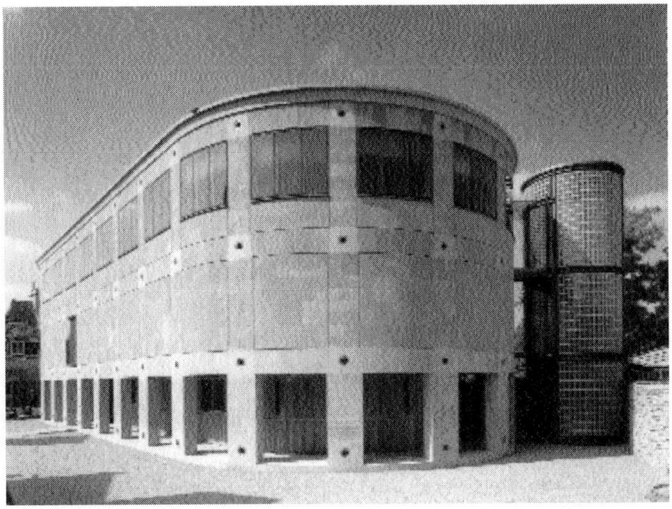

*Figure 4.16 Emmanuel College: external view – architect Michael Hopkins & Partners.*

chapel. Ketton stone is a machinable oolitic limestone (much like Bath stone), about whose physical properties little was known.

The conventional answer could have been a reinforced concrete frame, clad in thin stone panels with a steel trussed roof, fully air-conditioned and using ceiling forms to adjust the acoustic performance. Simplistically this involves quarrying the limestone, grinding it down, and firing it to create cement, which is then added to aggregate and reinforcing bars to create a structural frame and so on. What the Queen's Building did was to create insulation and mass by using the CAD possibilities now available for a construction in individual solid stones. Also it utilizes this thermal mass to reduce peak air-conditioning loads, plant and duct sizes as well as improve acoustic isolation. From a first cost point of view this might not have been the minimum option because of the intensity of labour required by the unusual construction process, but because of longevity (100 years design life) the possibility for reuse and lower energy running costs, the building definitely offers a sustainable design.

Because 'mass' gave considerable benefit to acoustic isolation from the nearby bus station the solid masonry carapace (with triple glazed windows) was conceived to support the heavy 'insulating roof' and perimeter loads from the interior reinforced concrete slabs. Mass also gave benefit to achieving the closely conditioned internal spaces where an acoustic performance of NR20 was required. The nature of the design process required adoption of the traditional procurement process using the flexibility of management contracting. 700 cubic metres of free stone was ordered by the client immediately on completion of the scheme design. This meant that design, analysis and testing of the project, organized with the help of the management contractor, all had to proceed in parallel with the procurement. Fabrication of the stone block, detail design and construction planning was itself an 'engineered' process. Testing, mostly at the labs of the Engineering Department of Cambridge University, involved detailed examination of small and larger sample columns to investigate the properties of creep and strength characteristic of the stone, in order to choose the correct mortar (8:8:1 lime sand cement), which would be crucial to the practical success of the stone frame. Detailed structural analysis of the stone frame proceeded in parallel (Figure 4.17).

The stone frame (without the traditional finials of a cathedral) achieves lateral stability by post tensioning with three interlinked vertical 32 cm diameter stainless steel Mac Alloy bars whose details were developed both to guarantee the client's aspirations of a building life of at least 100 years, and to reinforce the architecture as well as the engineering of this modern building in its traditional setting. These details included the corbel for the American oak and stainless steel roof structure, details of the shapes at the kneeler blocks and the construction details for the heavy roof cladding of oak panelling hung off stainless steel tubes. Internal lean mix pugging and external lead cladding

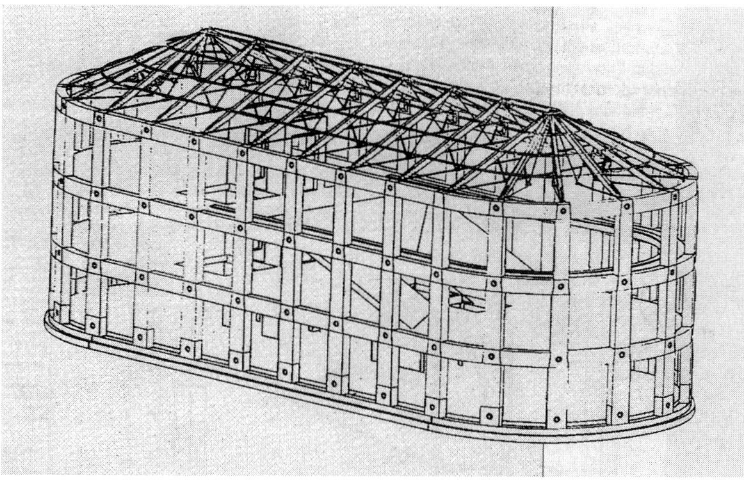

*Figure 4.17 Emmanuel College: three-dimensional model of the structural frame.*

added to the acoustic isolation by achieving a 2.5 $kN/m^2$ roof construction (Dickson and Werran, 1999).

Significant in achieving the architectural and engineering aims of this project was the transfer of the extensive craft knowledge of the mason Andre Vrona of Ketton Architectural Stone to bear on the more 'Classical' approach of the consultant team. This procurement also required an unusually interactive process between all the disciplines, as is required to effect most 'sustainable construction'. Particular knowledge and skills contributed to the common pot were:

| | |
|---|---|
| Satisfying a heritage environment | Client/planner |
| Vision and implementation of a timeless design | Architects/engineers |
| Creation of a performance space | Acoustician/AV consultant |
| Organization of the procurement process | Contract managers |
| Perfection of the craft of masonry | Stone masons |
| Advancement of the technology of stone framing | Structural engineers/ researchers |
| Opportunities for creating the internal environment | Environmental engineers |
| Costings specification of work packages and so on | Cost consultants, estimators |
| Construction of the works | Builders, fitters and others |
| Fitting out and furnishing | Carpet layers, cabinet makers, plasterers, electricians |
| Operation of the building | Fellows, staff, students and others |

Such a process is actually embodied formally in that of Concurrent Engineering within the Industrial Construction Process of Partnering – following the Egan Principles of *Rethinking Construction*.

### Hooke Park, Dorset

Designs for the School for Woodland Industries at Hooke Park, Dorset, first with Richard Burton of ABK on the prototype house and a 45 × 15 metre workshop building (Figure 4.18), and more recently with Edward Cullinan Architects on the eight rooms and seminar accommodation at Hooke Park, Westminster Lodge (Figure 4.19), have all aimed at developing an engineering and architecture exploiting Norwegian spruce thinnings, a product of the Hooke Wood itself. The workshop exploits

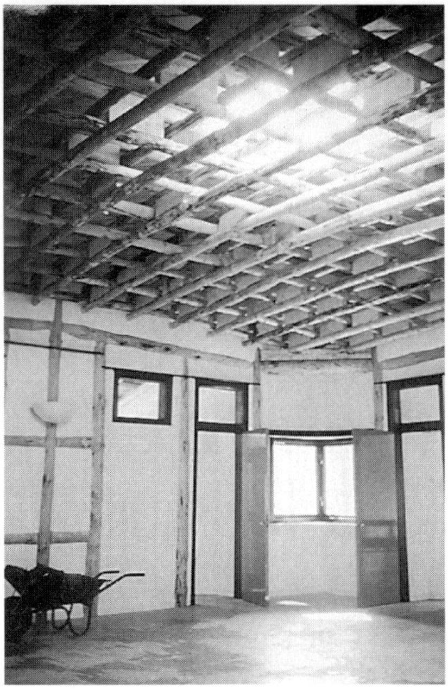

*Figure 4.18 Interior of Westminster Lodge at Hooke Park – architect ABK.*

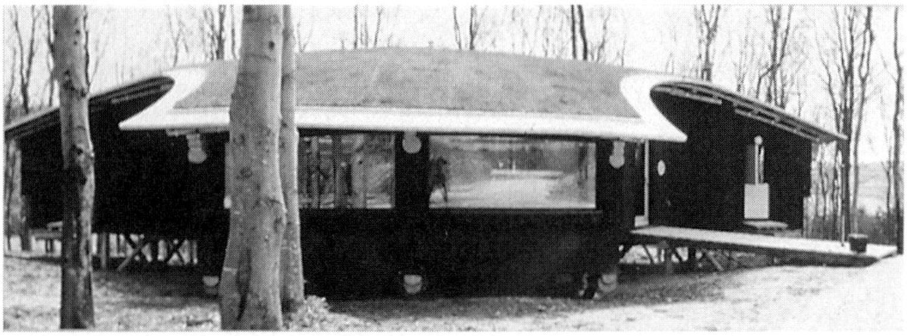

*Figure 4.19 Westminster Lodge – architect Edward Cullinan Architects.*

the flexibility of greenwood to form arches and transform these into a shell structure when natural lighting and ventilation are possible (Burton *et al.*, 1998).

The architecture of the lodge achieves its cruciform plan from the realization that accommodation is really about walls, floors and roofs. The cruciform plan with four sets of 3 × 6 metre rooms around a 9 metre seminar room also required a turf roof to ease planning acceptance of an additional building within Hooke Forest.

The subframe of walls and floors uses 150–175 mm diameter roundwood logs propped from single pad foundations by racking struts and sliding wedges. These walls were subsequently infilled with a double layer of plasterboard. The roof uses a composite double layer timber grid of continuously scarf jointed and glued members 90–110 mm in diameter which could otherwise only have been used for fence posts. These are formed into a shallowly curved double composite layer timber grid to support the turf roof (and students) over the 9 metre seminar space. The development of the glued scarf joint enabled the use of these thinnings and of the wedged raking joint and the engineering devices that determine the aesthetics of the final building and its treatment of the woodland floor.

## Building physics – a holistic approach towards sustainability[2]

### The BRE Energy Efficient Office of the Future

The BRE Energy Efficient Office of the Future (Figures 4.20 and 4.21) achieves its 40 out of 42 credits on the previous BREEAM rating and a

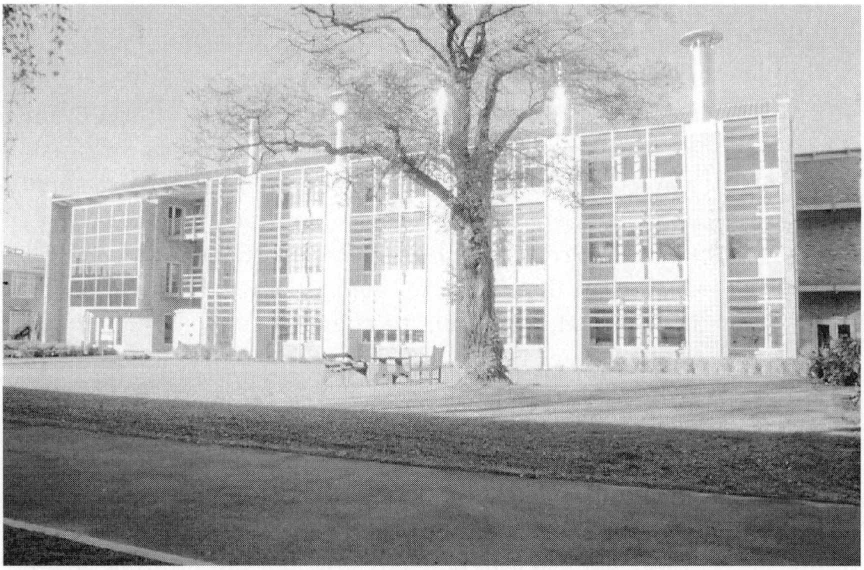

*Figure 4.20 Exterior of New Building 16 at BRE – architects Feilden Clegg.*

2 An approach anticipated by Banham (1969).

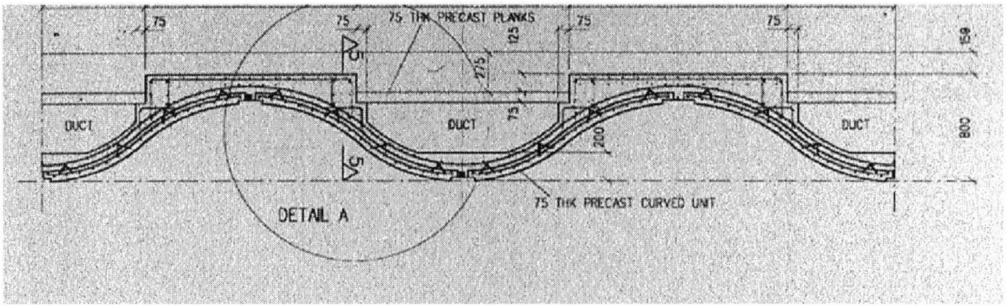


*Figure 4.21 New Building 16 at BRE – sectional view through floor.*

yearly performance of 80 kWh/$m^2$ by a radical approach to the engineering architecture which is expressed in the overall aesthetic of the building.

The potential problem of restricted cross-ventilation in cellular offices was addressed by incorporating floor ducts to supply fresh air deep into the floor plate capable of managing the large volumes of air. As a result of the collaboration between the architects, environmental engineers, Max Fordham Associates and the structural engineers, air is drawn through the bottom of the wave while the top of the wave allows daylight to penetrate deep into the office floor plate and artificial light to be reflected. The wave form is itself an efficient shell structure which minimizes the amount of material used and acts additionally as thermal mass. Construction is eased and the quality of finish is ensured by employing thin precast concrete unit forms whose eventual diaphragm action and acoustic and fire separation are obtained by *in situ* concrete topping. Precast concrete panels incorporated into the top of the air duct in the floor allow for the inclusion of embedded pipes cast into the floor screed, which can be cooled by deep borehole water.

**Learning Resource Centre, Thames Valley University**

In contrast, the Learning Resources Centre (1996) at Thames Valley University (Figure 4.22) benefits environmentally from a different flexible approach without loss of aesthetic originality, but relies for efficiency on machinery. Glare-free natural lighting, low-energy artificial lamps and a choice of tempered or cooled (minimal) mechanical ventilation with natural fan-assisted or tempered ventilation that varies with the seasons maintains minimum winter and maximum summer temperatures at 19–25 °C respectively. Treating each season differently by the adoption of different building services engineering techniques enables

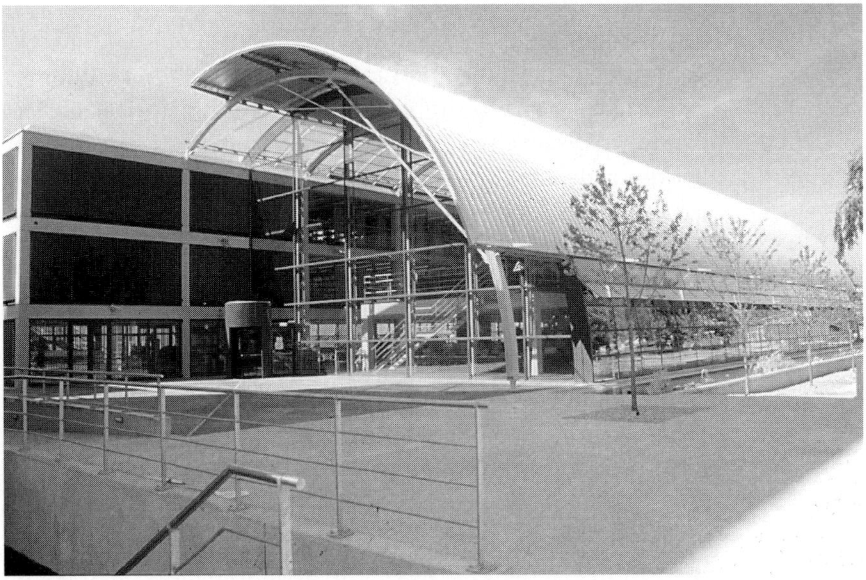

*Figure 4.22 Thames Valley University: external view – architect Richard Rogers Partnership.*

this building to achieve an energy consumption of 100 $kWh/m^2$ over a normal year without loss of architectural modernity.

### Tannahill Centre, Paisley

'Sustainability' can also mean the support which engineering and architecture can bring to the social well-being of a community. The £6.5m Tannahill Centre, Paisley, provides all the social facilities associated with a community function – GP surgery, family centre, community forum, housing association and management suite – together with shops and a post office. It was built with the help of Regional and European regeneration funds to help the Ferguslie community achieve a greater level of 'sustainability'. By design, the architecture of this protected family centre is mainly naturally lit and ventilated and uses a careful selection of materials to meet a restricted budget, while minimizing running costs. The atrium connecting the various functions uses an innovative translucent triple skin vector foil cushion roof to control solar gain, but admits natural daylighting while providing the insulation of the cushioned air to meet building control standards, at a fraction of the cost and weight of a triple glazing system.

## Summary

To gain 'Brundtland' points, clients, designers, constructors and users need to go back to the fundamentals of performance and examine the standards that have been set by the various codes of practice to which works are constructed. These need to be formulated on the basis of performance requirements rather than being purely prescriptive. Additionally, the following criteria among others are relevant.

- Occupiers are really integrators of a range of stimulations (aesthetic, thermal, acoustic, light and so on) and are more tolerant than anticipated by rigidly defined, legally set criteria of performance.
- Studying of solutions from the past together with the use of computing power and modern technology to evolve a new language of the built environment.
- Costing needs to be in terms of a total process of activity, adaptability, maintenance and deconstruction.
- Designs shaped to reflect the available materials, what types of physical and environmental forces they will be subject to and the socioeconomic culture of the 'local' construction industry and use of 'appropriate' technology.
- A conscious evaluation of embodied energy within construction (and the non-renewable materials required) in order to create a built environment which during its lifetime is effective and reduces the global requirement for non-renewable energy.
- Minimization of waste by use, recovery, reuse and recycling of materials.

The culture of our society is fast becoming more conscious of the sustainable. Driven at both government and personal levels, users, developers and designers are turning more and more to building solutions which drive down use of energy. Today's most advanced buildings, as a result, are fast becoming the collaborative efforts of professionals experiencing a revolution in the traditional ways that buildings are both procured and designed. The long-established boundaries between architecture, engineering and construction are dissolving to the mutual education and enlightenment of all professions. Designs emerging now are the result of true cross-discipline collaboration, based on the ever developing support and resources that technology can give and the refinement that industry can bring. This is the way to realize the vision of a client, beginning at the feasibility stage in order to achieve buildings that really do provide the right environment for use. Placed in the correct location we must recall that an urban building refurbishment project can be a greener thing to do than a new building on the edge of town.

As a Visiting Professor at the University of Bath, I see young people entering our profession with a degree in either architecture, engineering or construction, but having studied on a course where the disciplines are jointly taught. These are the new creative forces of tomorrow's professionals trained to see the whole picture, not just one part of it. The interface between engineering and architecture only works if engineers understand what architects do, architects understand what engineers do, and both understand what constructors do.

'Green' architecture is not just a style for altruistic developers to adopt. It is a movement gathering pace drawn by compelling economic and global reasons. The issues are very complex and different for each building usage, so that the language of sustainable architecture, engineering and construction for the built environment needs augmenting immediately, otherwise our activities as a species will be likely to overload the planet.

I believe that the methodology and examples given in this chapter show that the emerging architecture for our third millennial society is holistic in many ways, embracing equally environmental design, sociological and aesthetic factors. These driving factors will produce the new building aesthetic of the new millennium, where the materials usage, energy performance and use of a building play a key role on its visual and environmental impact.

## References

Banham, R. (1969) *The Architecture of the Well-Tempered Environment*. Architectural Press, London.

Barnes, M. and Dickson, M. (2000) *Wide Span Roof Structures*. Thomas Telford, London.

Burton, R., Dickson, M. and Harris, R. (1998) The use of roundwood thinnings in buildings – a case study. *International Journal of Research, Development, Demonstration and Innovation*, **26**(2), March/April.

DETR (1998) *Rethinking Construction: the Report of the Construction Task Force*, July 1998.

Dickson, M. and Werran, G. (1999) The post tensioned, prestressed Ketton Stone perimeter frame of the Queen's Building, Emmanuel College, Cambridge. *Structural Engineer*, **77**(20), October.

Dickson, M. (2000) Structures for the future – the search for quality developments in structural form to minimise environmental impact, *IABSE Symposium*, Rio de Janeiro, Brazil.

Groak, S. (1992) *The Idea of Building: Thought and Action in the Design and Production of Buildings*. E. & F. N. Spon, London.

Institution of Structural Engineers (1999) *Building a Sustainable Future – Construction without Depletion*. Institution of Structural Engineers, London.

Latham, M. (1994) *Constructing the Team. Final Report of the Government/Industry Review of Procurement and Contractual Arrangements in the UK Construction Industry*. HMSO, July 1994.

Markham, S. F. (1942) *Climate and the Energy of Nations*, 2nd edn. Oxford University Press, London.

Missenard, A. (1957) *In Search of Man* (English translation). Hawthorn Books, New York.

Pirsig, R. M. (1974) *Zen and the Art of Motorcycle Maintenance*. Morrow, New York.

Rice, P. (1994) *An Engineer Imagines*. Ellipsis/Artemis, London.

Walker, D. and Addis, B. (1997) *Happold – the Confidence to Build*. Happold Trust Publications, Bristol.

Watts, G. (1998) *A Reflection of Ten Years of the CIC 1988–1998*. Construction Industry Council, London.

## CHAPTER FIVE

# Synthetic thinking between engineers, architects and designers

*Ian Ritchie*

### Introduction

I want to start this chapter by affirming my own personal indebtedness to engineers; but I aim at the same time to help proclaim to a wider audience that art is an essential ingredient of all good design, although that art has been, and still is, so often buried beneath an avalanche of the architectural taste of the day. It was Buckminster Fuller who once said that most architects are simply good (or bad) taste-purchasing agents (that is, they are experienced at selecting from catalogues).

To enjoy the company of engineers, economists and those in the construction industry is a privilege, but for me they must be more than just concerned with their own world. They must be sensitive human beings, who have a personal philosophy about life in general. This is often too rare, or rarely apparent, but I have been fortunate to have met and worked with a few, and there is no doubt in my mind that they deserve public recognition (even if they do not seek it) on an equal basis to those architects they have worked with. What these people have in common is confidence, and an intuitive sense of sharing an experience where *the* job is the goal, not *their* job.

Numbers 'are not a sadistic conspiracy devised by one half of society for torturing the other half. Instead they are evidence of our inventive genius, and understanding them is one of the most important characteristics that distinguish us (as animals) from other animals' (McLeish, 1991).

If we add culture to numerical literacy, then we really do create the medium of invention. Today, this culture, at a personal level between individuals, can often begin with estranged collaborators, all of whom have to be capable of shedding their protective skins to enable a barrier-free exchange to take place – this is the primary move to allow cultural fertilization to take place.

This approach is no different in a larger context. History is full of nations and peoples developing philosophies and techniques which have, through mixing, produced the context for extraordinary creativity.

An exhibition is essentially about cultural exchange – the Royal Academy is such a venue. In 1992, an exhibition, *Art of the Structural Engineer*, had the themes of concept, form, materials, connection and construction – these are no different from the essential components of architecture, sculpture and even music, which is why there should and must be common ground between those of us who are fortunate to be entrusted with a major role in creating our built environment.

The true artist searches for the essence in things and then seeks to express it. It may be in the properties of a material, or how one part interacts with another. Engineering is a domain which should seek essence. Engineering is much more than a quiet service industry to architectural egos, but perhaps it is too provocative to suggest that, without engineers, architects would collapse.

## Synthetic thinking

The profession of civil engineering and its offspring, structural engineering, has its roots in the discipline and hierarchical nature of imperial armies acting on behalf of certain Western countries. Here were born the tenets of economy and efficiency.

Economy and efficiency are the historic buzzwords of engineering design for the traditional engineer. When used by visually illiterate engineers as the only design criteria, they have led to a great number of 'aesthetically' unsatisfactory structures. However, it is the attitude and lack of design skills rather than these tenets themselves which have produced these results. There is no reason to suppose that we cannot make economy and efficiency subservient, without denying their crucial importance in the design process and eventual artefact.

During this 'colonial' period the architect became the engineering decorator, with a role to camouflage – arguably carried out in the interest of urbanity, architecture or simply making the engineering publicly acceptable (for example Tower Bridge). This division has existed for nearly two centuries. There have been enlightened engineers (Brunel, Telford, Eiffel, Nervi) who have produced exceptional engineering works which are not only structurally inventive, but go beyond the material domain to create notable public space. These engineers clearly had an awareness of the civic importance of some of their works.

Architects who have recently been in the vanguard of structural inventiveness in their architecture have been so only because of the support of engineers, yet the public's appreciation of the engineer has been severely limited by the media's sole promotion of the architect.

In the late 20th century, structural engineering inventiveness now requires the support of rigorous analytical methods. It has not always been so, but we have experienced an ever-increasing tendency towards this position as a result of material research, technical development and the application of new materials. Architecture is as much a witness to this as is engineering.

Masonry, as the main engineering material, marked most of mankind's achievements up to the 19th century. Then, with the advent of wrought iron (the 30 m span Coalbrookedale Bridge by Abraham Derby, 1799), then wrought plus cast iron (Thomas Telford's 177 m Menai Strait Suspension Bridge, 1820; Paxton's Crystal Palace, 1851), then iron plus steel (Gustaf Eiffel's work, notably the 178 m steel arch of the Garabit Viaduct, 1884) and finally steel in cable form, structural engineering with steel has become more and more specialized. In the context of light and architecture, the extraordinary achievements of Paxton and others in the mid-19th century in removing the entire solidity–opacity associated with buildings, has resulted in a renaissance in today's architecture (Figure 5.1).

The innovation of reinforced concrete led to many advances – notably by Auguste Perret and Eugene Freyssinet in France; and in Switzerland by Robert Maillart, who, in 1901, designed the 38 m span hollow box arch form Zuoz Bridge. This material has seen tremendous advances since, through a better and better understanding of the nature and quality of the materials involved and how they behave together. The engineer can now bring to this knowledge yet more powerful analytical models to test ever more complex solutions.

There have always been individuals who, either ignoring or defying professional boundaries were capable of pan-professional action (for example Eiffel, Nervi and Fuller). Of course, this behaviour attracts envy and creates jealousy from traditional professionals. It is rare today for an individual 'engineer–architect–designer' to be capable of thoroughly analyzing inventive structural engineering proposals without the support of specialists. Such a person may appear to have a broader perspective within which to design, but this can be less than the horizons imagined by a collaborative team of creative individuals who come from different disciplines (for example Rice Francis Ritchie), yet who can also continue to develop individually within their own field.

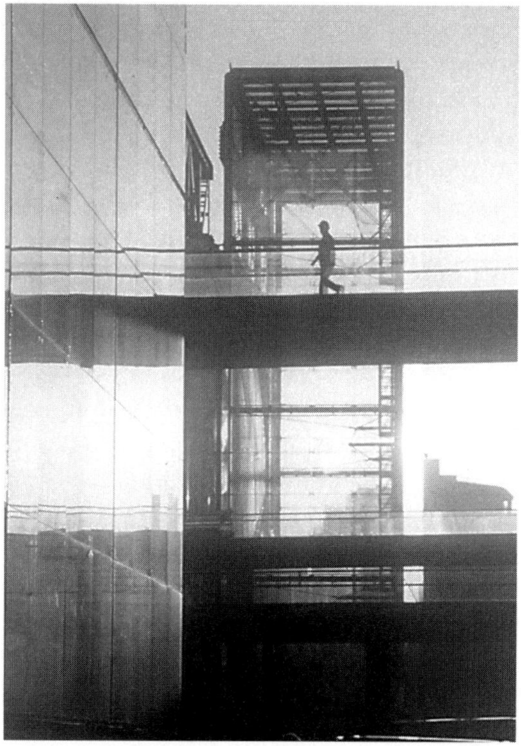

*Figure 5.1 La Villette: Façades Bioclimatiques – (RFR). Rice Francis Ritchie (RFR) created the architectural and engineering design of the three grand bioclimatic façades of the new Cité des Sciences et de l'Industrie in Paris. The design was based upon the competition-winning museum architecture proposal of Adrien Fainsilber. These three façades are key aspects of the building's architecture. They act as intermediate zones between the outside and the inside of the building, filtering light and energy and acting as huge windows seen through from both the building and the park. The interpretation of transparency became the key architectural idea in the development of the design. Secondary concerns were ideas of tension and the didactic role that the facades could play as part of the Cité des Sciences. The major innovation, initiated by the desire to achieve and define a material transparency was in the structural exploitation of glass to enhance the perception of transparency.*

Imagination–creativity–intuition, material understanding–analysis–design–economics, and an understanding of the political and social role of the process through which we realize projects are all ingredients in the making of architecture, and dissolving the intellectual boundaries between professionals is a fundamental necessity if we are to realize more intelligent and responsive architecture.

## Professional barriers

Psychological barriers only occur in people's minds, and like any theory constructed by people these barriers can be deconstructed and replaced. 'Barrier absence' requires a way of thinking and attitude which is no longer territorial, since respect and trust exist, qualities which in turn encourage confidence with humility between people. Professionals should be as capable of realizing this as anyone else, and in terms of their influence on society and the physical environment should have a moral obligation to do so.

I know from my own experience with Peter Rice and Martin Francis, and the way our office in London functions with individual engineers and economists, that territories do not have boundaries: they are simply different landscapes which require different skills to negotiate well, but also through which, with one's collaborators, one can be supported and supportive.

In an architectural practice which has at various times included engineer, naval architect, artist, anthropologist, photographer, landscape architect and poet, where none are regarded as technicians or draughting people, we are accustomed to the absence of barriers as we are to the absence of hierarchy. Since a small office cannot always sustain this diversity, we frequently collaborate with other professions in the usual way. It is important to do this with people who share the same objectives – for example design quality and similar values. Thus we have, over the years, established working relationships with a selected number of consultants.

Mutual education and reorientation are necessary when a job comes to us with another consultant already attached to it by the client, and there is a 'heat' period necessary to melt the engineering and architectural boundaries.

## Collaboration

The kernel of creative collaboration contains several crucial ingredients.

- Each must take time to listen to the other, and suspend prejudices. Not only does this allow mutual respect to grow, but without it, the synergy of mutual creativity cannot flourish. The process is like brainstorming, in which nobody can quite remember where the solution came from.
- The commonality of aims is usefully complemented by a diversity of expertise.

- No barriers = no defences. There are those who feel threatened when another profession speaks their language and questions their assumption. It is a shame we have these languages and hide behind them at times. This is inhibiting to any free exchange.

These principles of collaboration apply at all stages of a project: initially with a client and consultants, and later with a builder or fabricator. Differences of orientation can generate conflict, or can be harnessed creatively. The trick to help this process move in a constructive direction is often found by sticking rigorously to an open-ended approach where everyone's preconceptions – especially our own – are questioned, and we demonstrate a willingness to receive others' ideas and modify our own, while at the same time refusing to compromise our design principles and values. Those ideas that survive this process of challenge are the stronger for having stood up to scrutiny, and the process is exciting.

> *After all, we all agree on that [collaboration]. But talking about it doesn't seem to have had much effect. One must somehow create the conditions which will allow such collaboration to take place, and one must educate members of the building team to see their own contribution not as an end in itself, but as a part of a common endeavour to create comprehensive, total architecture. (Arup, 1972)*

Here are Ten Commandments for Collaboration:

1. There has to be a moral commitment.
2. There should be no preconceived idea and collaborators should be open to almost anything.
3. Learn to really listen and to interrupt, and be ready to be interrupted.
4. Ideas are shared – no one can claim them afterwards.
5. Be altruistic, not competitive.
6. Respect the minds of your collaborators; their individual skills will become valuable later.
7. There is time together – synthetic time – and time alone – reflective time.
8. All participants are equal; there are no bosses.
9. You have to respect the common concept as being more important than one you could have conceived by yourself.
10. Be prepared to improvise.

All of our work has a public content, whether the entire building or just its facades. We rarely, if ever, construct with our own money for ourselves. We act as the group mediating between the public and the client, and together we have an obligation to both. When we collaborate, some degree of friction always arises, whether it is over the money, the design or the morals, and has to be resolved in the end by the project itself. It is recognizing the project as mediator which helps to solve it.

The collective idea which emerges is the first and most important thing. It is vital that all who are to collaborate on the design of a project come together at the beginning. All design work is political in the ultimate sense in that we are trying to produce a world that is better to live in, where people understand more, where people are less oppressed and people live less dreary lives, and where they have more control over their environment: this is the glue that brings us together.

The person who first receives a call from a client has to decide very quickly how and with whom he or she will work. It is often an architect or an engineer. Such people make up their own mind, but also have some obligation, as soon as they set up this wheel of collaboration, to inform the client. The defining of issues between collaborators, such as the nature of the contract, fee split, joint or separate insurance and responsibilities, is very important to pin down early. Collaboration does not really mean anything until it is defined.

Some of the best concepts have come from people who are not 'recognized designers', yet who are as concerned about our future environment as architects, engineers and landscape designers. These include geographers, urban planners, archaeologists, anthropologists, artists and poets. These people understand how environments work as well as architects. For me, open collaboration, where individual egos work together in the interest of the project and beyond, is the most enjoyable method and approach. Shotgun weddings are very difficult to manage.

## An open methodology

I have tried to outline a methodology, a way of working, which can engage different kinds of people. For I think the world of tomorrow will have problems that are more complex. Architects, engineers, designers and artists will have to find new ways of solving them with their clients. We will all have to learn new kinds of methodologies to solve problems. We need to approach each project with fresh eyes and minds to seek, with the client, the best way to achieve the desired end result. Traditional forms of engagement, of construction contract, of collaboration, are constantly being questioned by us.

## Synthetic thinking in a more complex world

While our designs have often explored the structural and energy performance of certain materials to help create spatial environments, one of our

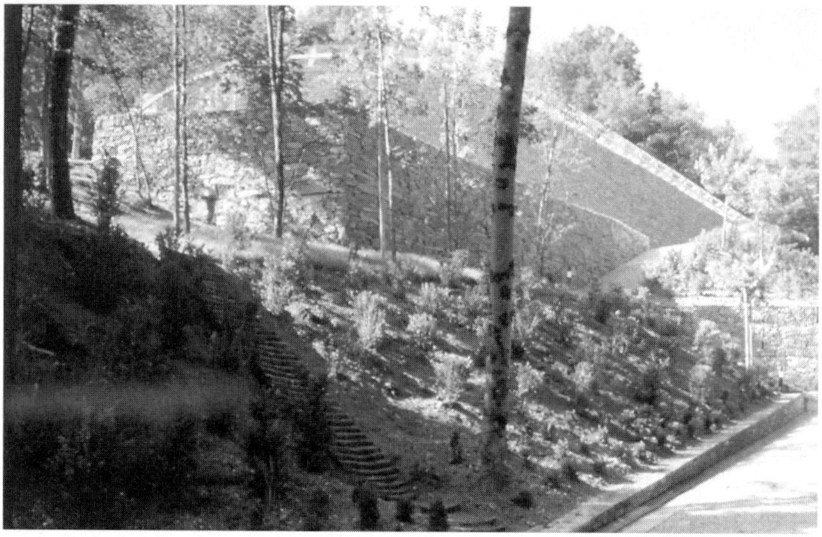

*Figure 5.2 Terrasson Cultural Greenhouse (Ian Ritchie Architects, Kathryn Gustafson, Ove Arup & Partners). The building, designed in 1992, was conceived as a public performance space for theatre, conferences, exhibitions and other municipal events, as well as having the potential to incorporate a reference library and public research centre on plants. It is a peaceful sheltered space. The greenhouse leans symbolically against the hill. Its clear glass flat roof reflects, like a lake, the changing sky and the foliage of the surrounding trees. The steep incline of the site and the need to retain the soil in the garden led to the idea of using 'gabion' – steel mesh cages holding stones – for the greenhouse walls. The greenhouse gabions are freestanding vertical cantilevers, and allow the building to breathe naturally. Internally, the gabion wall is lined by citrus trees. The building enclosure creates its own internal micro-climate throughout the year. The concept sought to highlight the contrast between a highly processed and sophisticated building material and the unprocessed stone from the local quarry. The use of gabions as a primary building element is thought to be the world's first such application. The building and the new 5 hectare park and the greenhouse opened to the general public in spring 1996.*

current concerns, as illustrated by the experimental greenhouse in Terrasson (Figure 5.2) is to create a less expensive architecture using material which is less and less processed by industry, while maintaining the pleasure of light in architecture. At the same time, we stay aware of research and developments in what I refer to as high technology, such as holographic films to help conduct and distribute light using far less transparent surface areas; molecular 'replicating' spider's webs; and nanotechnology (advancing in Saarbrücken at the Institute for New Materials, and elsewhere).

At the end of the 20th century we recognized some of the environmental building follies of the century, such as air-conditioning. We can also see the crudeness of our industrial manipulation of the Earth's resources into fairly basic building materials.

Looking ahead, I can see much cause for optimism. One need only investigate a tree. Trees capture light, make energy, grow by processing $CO_2$ and water, and support and are a home to other life forms. They don't make a humanly audible sound, they don't appear to waste heat or energy (or waste anything else for that matter) and are natural pollutant processors. They provide shade and they look great. If there are appropriate architectural forms, they are to be found in nature (Figure 5.3), but we need to understand their composition and mechanisms at the molecular level, not simply appropriate their forms for visual delight.

I would like to quote an exchange from the second reading of the Energy Conservation Bill:

> *(Mr Patrick Thompson) ... Concepts such as wind power, wave power, hot rocks and fuel cells are exciting, but, as my Noble Friend said, the truth about energy efficiency is that it is boring, as it really only concerns property insulation and design.*
>
> *(Mr Brandreth) I disagree with my Hon. Friend about energy efficiency being boring. I find that when I talk to people in my constituency about cost savings, they recognise that it is a case of enlightened self-interest and that they can help save the world, while improving the quality of their home life and enhancing their bank balance. Far from being boring, it is exciting.*

## The complex problems of tomorrow

A new, wider and more appreciative Europe will not, hopefully, be the only creation in the coming years of the largest, most powerful single economic market that the world has ever seen, with its consequent energy growth demands, but also a staging post, symbolic of the desire to achieve a more integrated whole world.

Monetary economics has so far failed to find a way of dealing with social costs or with renewable resources. The present Western mania (indeed more and more global) for development based on a mechanistic

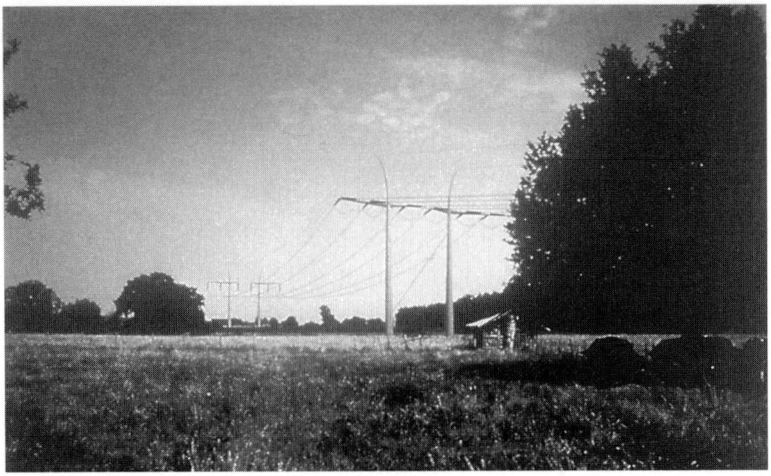

*Figure 5.3 EDF Pylons (Ian Ritchie Architects, Kathryn Gustafson, RFR). The team of Ian Ritchie Architects, RFR and Kathryn Gustafson were invited in 1995 by Electricité de France to compete against seven other teams in an international competition to design a new series of very high voltage pylons – 225 kV and 400 kV. The EDF awarded joint first prize to our team's proposal and to that of Marc Mimram. Our 400 kV 'f' pylon design has been developed and full-size prototypes successfully tested in 1998. The first of the new series will be installed in 2001 across the Rhône in Le Gard. EDF anticipates replacing and installing up to 4000 new pylons over the next ten years. Our concept was developed upon a philosophical investigation into the contemporary meaning of progress. The design concept is based upon a single-level configuration of the conductors which reduces the height of the pylons, producing a horizontal and discreet expression. The design aimed to establish a clear spatial relationship between the lines of the conductor cables and the landscape over which they pass. To do this we developed a flexible family of pylons. The pylon's structural form is an 'f' carrying a three-phase single circuit and is made of rolled steel plate. They are twinned to carry two circuits. The pylon column enters the ground with no detail. The EDF has named our 'f' pylon* fougère *(fern).*

and materialistic viewpoint, supported by the present inadequate economic methodology, has led to increased pollution on a global and local scale. Yet to most economists it appears that the social and environmental costs still remain intangible.

One may think that the point of economics is to help us manage the world better. However, I suspect that few economists see it this way. It seems inevitable that there must be a change in the current economic way of thinking. Humanity developed the present model, and our actions still maintain it. A sustainable economy means a more compassionate one, in the way we relate to each other and the planet. The Earth owes humans nothing. The global spread of the free market economy (so far leaving aside the polar regions) sucking up the Earth's wealth will probably lead our present concept of progress into oblivion. Excessive borrowing from each other and our children is wrong.

Exchange through discussion and openness of information is essential for understanding, and when this engages cultural exchange a major prerequisite for creativity is in place. This in turn makes creativity more accessible and possibly more democratic in a less competitive environment. Competition has been and remains the conceptual trigger of our present economy and society. We do not believe that this is inevitable, as is often argued. Collaboration, cooperation and indeed altruism are as common a natural inheritance as 'survival of the fittest'. This is a clue to redefining economic ethics, where the economy is seen not only to serve people in a material sense, but in a wider, more holistic context, where non-material issues are as important as material ones.

Quantum mechanics has shown that we are not observers but participants in the world around us, and yet through our limited human perceptions we continue to describe and prescribe as if our minds were still outside our own bodies and environment. The study of ecology has brought this into focus for us. Science has also shown us that the only certainty about certainty is uncertainty. These observations, which have undermined science's own imperious position, ought to be making us more tolerant and more ready to participate together in many more aspects of life.

Is not the art of living the ultimate art? In the end, it is not the planet which is at risk but humanity's place (and existence) upon it. Our present concern for the planet appears to be a reflection of our selfishness (for example the Energy Conservation Bill debate quoted above). We hear and talk about the loss of the world's natural resources, plant and living organisms, both in our own country and across the planet, but more often than not this is done in the camouflaged context of our human survival through nature's diverse resources for human welfare (medical and so on).

It is important to participate as an individual to express concerns about the wider issues which affect architecture. Like holography, in each part is the whole, but unlike the hologram, each part is important to the whole. Ideas popularized in the 1960s – dematerialism (conceptual art), ecological awareness, spaceship Earth, world citizenship and the revolution in lifestyles (challenging sexual and religious taboos) – have been slowly and discreetly absorbed and transformed by Western society into more practical and vociferous views on how to begin solving world issues such as hunger, pollution, shelter and intercultural communications. (Some would argue that this is potentially another form of colonialism.) Yet the economic model remains largely impervious to such ideas.

In our urban environments, also spreading globally and now 'home' to a larger and larger percentage of the Earth's inhabitants, there is a danger that we are establishing an exaggerated and cocooned sense of our own self-sufficiency, which in turn will further alienate us from the essence of life on Earth. Urban sprawl, now a major world environmental issue, was not even on the agenda of the world's first environmental conference in Rio de Janeiro in 1992.

The architecture we produce, and how we make our buildings, is a reflection of our world view, or 'how we walk on the Earth'. This is difficult for architects to assess in real terms, in a society still dominated by the culture of science and technology within the present economic model. Access to hard facts on energy, labour, social impact, recyclability and the renewability of materials used in construction is very difficult. Yet graphs depicting comparative energy consumption of, for example, extracting raw materials or of processing them do exist.

However, these 'facts', important as they are in signalling awareness, represent little in terms of the more complete picture. For example, we do not necessarily have the combined knowledge of the energy sources used, their comparative polluting effects, the effect of the production processes on the health of workers in these industries and their consequent social as well as economic cost and so on.

It is a mistake to assume that graphs and tables such as these can give a whole picture. Their importance and our dependence on such abstracted and limited data discredits us. It is in these sorts of areas that information needs careful examination, but will ultimately, one hopes, through significant development, begin to give us clear data on which to make our more holistic judgements on not just materials, but the entire construction and deconstruction process, and the way we access and use our built architecture; in fact, a more whole picture of the consequences of our decisions and choices.

Another important viewpoint is the effect (visual, psychological, physiological) of architecture on the 'user'. Healthy built environments for humans are one important aspect of a continuing biosphere.

There can be no sudden revolution in the industrialized regions of the world with regard to the way we extract, process, manufacture, distribute and consume materials. Only a long campaign will eventually change our habits, of which this book is a small but important part.

## Conclusion

Humanity and intelligence have as much to do with the process of decision-making as with the tangible artefacts which result from our application of science, technology and economics.

The need to make evident metaphorical intelligence and humanity in what we design should be indisputable. It is this which drives our design approach.

Real progress for mankind and a real sustainable future for the Earth are becoming essentially the same. Architectural and engineering design and construction must deal with their own progress by drawing upon the strong metaphorical stem of the human spirit and earthly values.

## References and further reading

Addis, W. (1994) *The Art of the Structural Engineer*. Ellipsis, London.

Arup, O. (1972) Lecture at the Institution of Civil Engineers, 26 October.

McLeish, J. (1991) *Number: From Cave People to Computers, a Revolutionary View of Ourselves*. Bloomsbury, London.

Ritchie, I. (1994) *(Well) Connected Architecture*. Academy (UK) and Ernst & Sohn (Germany).

Ritchie, I. (1997) *The Biggest Glass Palace in The World*. Ellipsis, London.

Rocca, A. (1999) *Ian Ritchie, Technoecologia*. Motta, Italy.

## CHAPTER SIX

# Different models of interdisciplinary collaboration

*Sam Price*

Before I turn to the main subject of this chapter, I want to make two general points. The first is that design is a means to an end, and the purpose of our efforts is to produce good *buildings*. We are designing things to be *made*. This should perhaps go without saying, but I think that in the introspective heart-searchings that sometimes go on about design the builders are quite often forgotten. To produce a good building, however 'good' is defined, whether it means beautiful or functional or economical or appropriate or honest, or preferably a happy combination of all of these virtues, one has to have an aim, an idea, which must be pursued single-mindedly, even with passion, and that passion has to be understood and shared by everyone involved and ultimately by the builders, who, after all, are the ones who create the building. 'Passion' may be an overstatement, and perhaps I am exaggerating in order to make the point, but I think it is often easy to see which of the buildings around us are the result of what I have called passion and which are the result of pragmatism.

I said that for the building to be good the passion has to be shared by the builders, by the craftsmen who are actually making it – laying the bricks, pouring the concrete, putting in the services – and I think it must be accepted that in recent years it has become gradually more and more difficult for the designers to communicate directly with the craftsmen. 'Builders' are becoming a thing of the past and most building work nowadays is organized by 'contractors', who do little or none of the work themselves and let it all out as a series of subcontracts. The contractors are trained in management and talk about 'procurement' and describe the building as the 'product'. None of this has much to do with passion. I doubt if passion is on the curriculum of many management courses!

So the designers are separated by contract administration and management from the people who will build the building and it is becoming progressively harder to get the message across, and this means that the message has to be clearer and more complete. What I mean by this is that the designers must have a stronger and clearer view of how the building will be built.

Now all this does have some relevance to the issue of interdisciplinary collaboration. The design must be convincing and complete whichever way you look at it – particularly from the builder's point of view. The different designers in the team bring different strengths and weaknesses to their work and it is at least possible that the engineers are the ones who are most inclined to be interested in the actual process of building. Engineering is, after all, all about making things – it is the practical side of science. Architects tend to be somewhat further off the ground – which is of course a very good thing – somebody has to be. It is the engineers who may have something to contribute to the realization of the gleam in the architect's eye, because that is where the gleam certainly has to start. So this is one of the great values of true interdisciplinary collaboration – that by working together we have a much better chance of achieving the completeness of design which is necessary if the passion is to get through to the people who actually make the building.

My second point concerns the nature of the interdisciplinary design team. I did for a time think that this complicated conglomeration of consultants was a very poor substitute for a single all-knowing designer who understood all the criteria for building. I thought that if we were all much better and more broadly educated we could each carry out the design without the need for help and advice from others. Education was clearly the key. There was in fact a conference two years ago on Design Education for Engineers at which the chairman, Lord St John of Fawsley, expressed this same view when he said in his introduction, 'Specialisation is the enemy. We should be seeking to rid ourselves of it. What we want is the Renaissance engineer ...'. I think that this may very well be the answer to the design of bridges, but I am now certain that I do not believe that it is the answer for the design of buildings. If we are to deal with the increasingly demanding environmental and energy-use questions and if we are to exploit fully the potential of the many materials that are now available to us we must have specialists who first and foremost are experts in their own field. No amount of enthusiasm for the general issues will make up for lack of ability in the specialism. Of course, nothing will be achieved by a bunch of experts who have no common ground and no common language, which is why the whole idea of interdisciplinary collaboration is vital to success. The other vital ingredient is enthusiasm, because without it we cannot achieve anything worthwhile, as I said at the beginning.

## Three methods of working

Now to the subject of this chapter – models of interdisciplinary collaboration. In thinking about what I wanted to say I decided that 'models' in this title could either describe the method of working – the setup in the design office or offices – or it could describe the building produced, as a 'model' of interdisciplinary working. So I have decided to describe and show you three examples: three different methods of working and three buildings which were produced by these methods and are all in their ways 'models' of interdisciplinary design. I will try to outline briefly the successes and failures, as I see them, in the buildings and in the teams.

## Ove Arup & Partners

My first example comes from the late 1960s, when I was working in Ove Arup & Partners. Arup had worked for some years with architects Sir Basil Spence, Bonnington and Collins. Jack Bonnington was keen to try interdisciplinary working, and had set up his own in-house interior design and services teams. He persuaded Reynolds and Young to put a team of quantity surveyors in his office, and I led a team of structural engineers from Arup. The organization of the office space was interesting. The architects, interior designers, services and structural engineers sat in one office space divided into bays by nearly full-height partitions, with each discipline occupying a bay. The partitions did not separate the spaces acoustically, so that we could talk to each other across the partitions, but there was a definite implication of separate departments. The quantity surveyors were, perhaps characteristically, more conservative and put themselves in a separate office with a front door and a knocker!

The main building that we were working on was the Sunderland Civic Centre (Figures 6.1–6.3), and the method of working seemed to me to be very successful. The architects, interior designers and structural engineers, particularly, got to know and understand each other very well, had lunch together frequently, went to Heals and argued about furniture design and light fittings and generally sparked ideas off each other. During the development of the design, which I will describe in a moment, everything went very well, and the job was generally very successful. It was only when the pressure came on from outside – the contractor pressing for information, for example – that the two firms tended, inevitably, to separate and take up defensive positions.

The main part of the building was conceived as a figure-of-eight, forming two irregular hexagonal courtyards based on a geometry of 60°. The strip of offices surrounding the courtyards would be either about 12

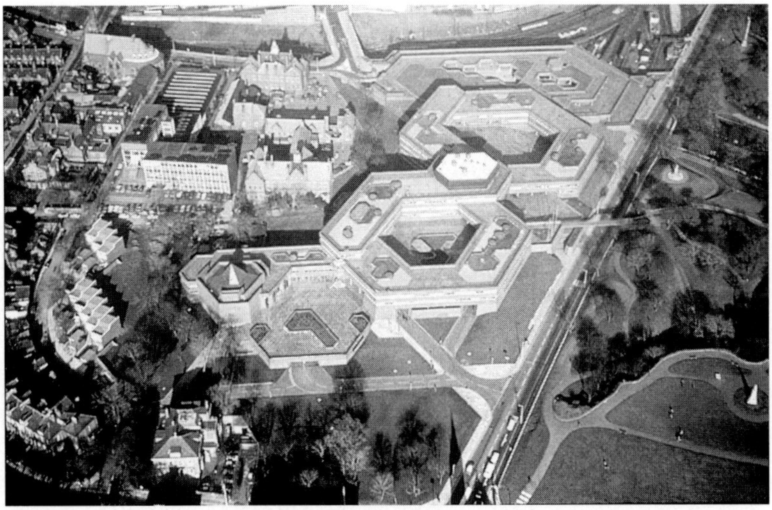

*Figure 6.1 Sunderland Civic Centre.*

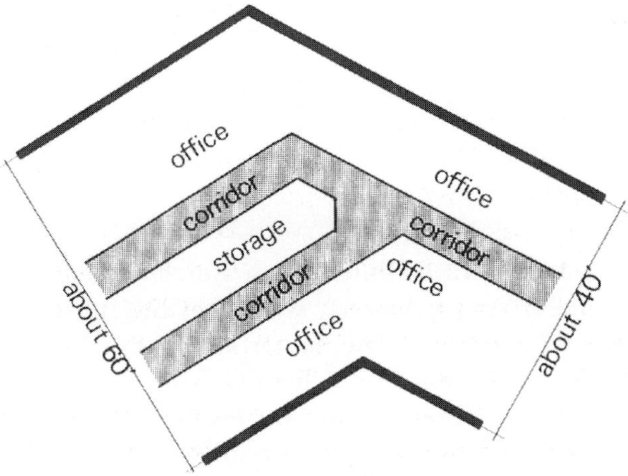


*Figure 6.2 Sunderland Civic Centre – the architect's initial scheme.*

metres wide, where there would be a central corridor with offices either side, or about 18 metres wide, where storage rooms or stairs or lavatories occurred in the middle between two corridors. This was the idea, and the question was how to structure it and deal with the interesting and unusual geometry.

I proposed that the whole building should be based on a regular triangular grid of 6.0 metre side equilateral triangles, broken down into 1.5 metre triangles, covering the whole site. The architect's preliminary plan for the whole building was, so to speak, picked up and shaken, and

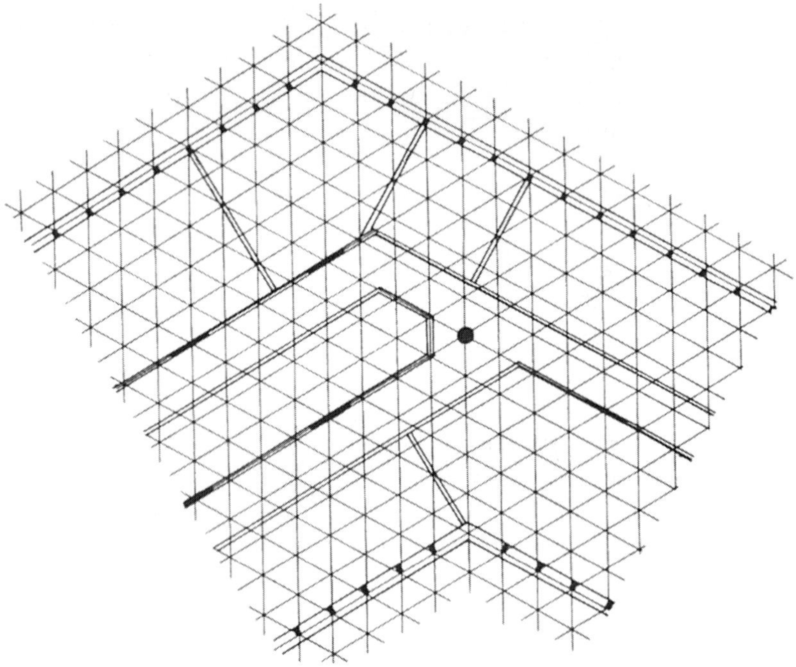

*Figure 6.3 Sunderland Civic Centre – the structural grid applied to the architect's scheme.*

with very little difficulty we then had a building plan based on the new underlying geometry, which set the rules for the development of the whole design. The material for the frame was clearly reinforced concrete; there were load-bearing mullions at the edge which were precast, because they were only 125 mm thick; all the rest of the concrete was cast *in situ*; and the floor slabs were 225 mm thick flat slabs. In due course, as the detail design progressed, we found that it was easiest and most economical to reinforce all the slabs with three-way reinforcement, so in every way the geometry of the design was followed.

I can't really pretend that this very effective and satisfactory coincidence of planning and structure could only have been achieved by the method of interdisciplinary collaboration that we adopted. Probably that fundamental breakthrough would have occurred with the structural engineers working in their own office. However, the subsequent detail design of all the elements – walls, stairs, cladding panels, services – was certainly made easier by the proximity of the architects and engineers.

At the end of the job I felt that it had been a most rewarding experience, but that what had happened on Sunderland – that the architects produced a scheme and the engineers found a system for achieving it – was probably as far as the collaboration would go. One might wonder what more could be expected – after all, the architects did adopt completely the engineer's proposal, and this is at least as good as most collaborations get. However, my appetite had been whetted and I wanted to see if collaboration could exist at a more fundamental level of design. So I asked to move to Arup Associates, which was a truly multidisciplinary practice with all the necessary experts in one firm.

## Arup Associates

Arup Associates was at that time organized into seven groups, each containing about 15–20 people. A typical setup would be four or five architects, two structural engineers, two mechanical engineers, one electrical engineer, one plumbing engineer, three quantity surveyors, maybe one building surveyor, an administrator and a secretary. The groups sat in open plan offices and usually not in professional groups. Note that there were no draughtsmen, which was unusual for an engineer's office at that time. This was because it was felt that the group should be as small as possible to keep internal communications crisp and to help the whole group to focus on a common aim. There were frequent discussions within the groups about every aspect of the design. Everyone's opinion on any subject was valid – if the plumber said that he didn't like the way the architect was designing the windows, then the architect had to be prepared to discuss it and explain why he wanted them like that. If you went into another group and talked to the mechanical engineer he would be quite likely to start to explain the structural principles of the building he was working on.

In many ways this system was fantastic. First, of course, for the client, who only had one firm to deal with and was never faced with buck-passing between different offices. Secondly for overall commitment to things like programme and budget. Thirdly because it allowed people to do what they wanted – for example the contract administration was all carried out by the quantity surveyors, so that the architects were free to concentrate on design.

During the nine or so years that I worked in Arup Associates we designed and built a number of interesting buildings. I think that nearly all of them demonstrate a fundamental interest in construction, and do celebrate the materials they are made of, and did try to give the craftsmen the scope to show what they could do. In other words I think that you can see the idea and the passion coming through. Perhaps the most interesting is the Lloyds Building at Chatham, which is a good example of the influence of group working at every level of the design.

### Lloyd's Chatham

Lloyd's wanted an office of about 20 000 square metres and a big underground computer room on the old Gun Wharf at Chatham. The site is on the banks of the Medway and is on three levels, separated by quite high retaining walls, the lowest level being the quay beside the river (Figure 6.4). The principal architect of the group proposed a very simple cube of building standing on the middle level, rather like the way that Aalto's Enzo Gutzeit building overlooks the harbour at Helsinki, and fully air-conditioned. Now this was in the mid-1970s at the time of the oil crisis, and we were all becoming very conscious of the amount of energy used by buildings. There was a general discussion in the group about fundamentals. The site was large and the air was clean and there were no obvious reasons why a narrow, naturally ventilated building could not be got on the site. The group generally felt that the architect should reconsider his starting point. I think this view was expressed most forcibly by the structural engineers, who always tended to be the stroppiest. The architect not surprisingly took umbrage. (This is certainly an example of one of the problems of group working – it can be very

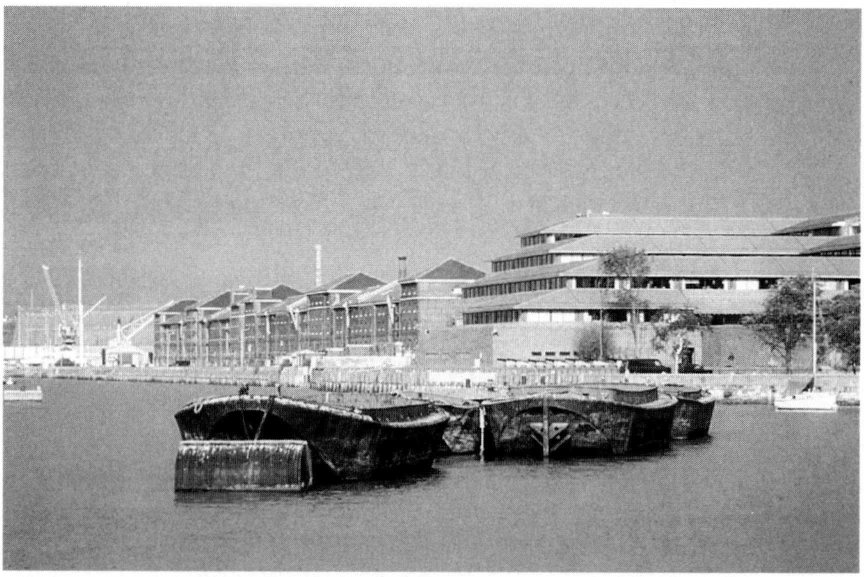

*Figure 6.4 Lloyd's building, Chatham – view from the Medway.*

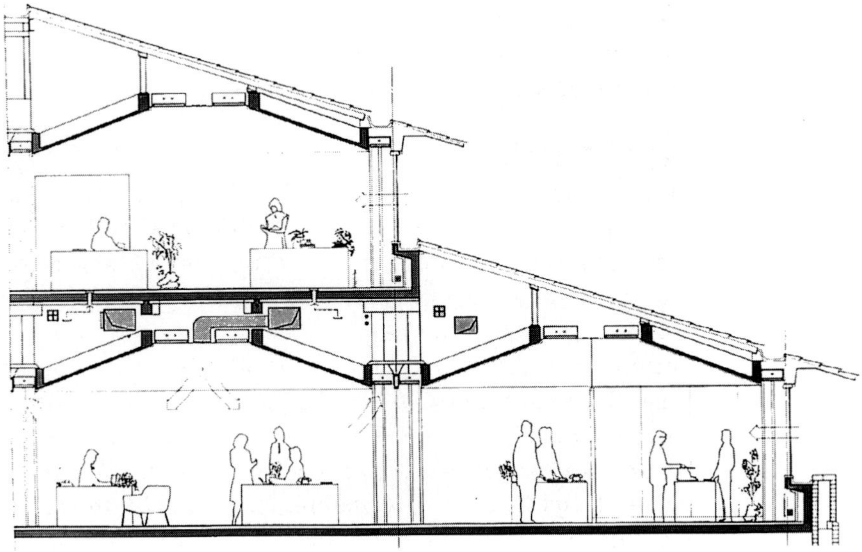

*Figure 6.5 Lloyd's building, Chatham – section.*

painful.) He invited the structural engineers to do their own scheme if that was how they felt.

So one of the engineers produced a sketch for a building, narrow enough to be naturally ventilated, with two open courtyards. It was a pretty scruffy effort and certainly lacked any architectural quality, but it did make the point that the idea was feasible. The architect, to do him justice, did pick up the sketch and turned it into a presentable proposition (Figure 6.5). This little episode demonstrates just how far Arup Associates managed to get in developing the democracy of the group.

The design of the building developed – following very closely Philip Dowson's principle of 'Constructional Honesty'. We wanted a square grid with vaulted ceilings to give a slow rhythm and a sense of place in big open-plan offices; no false ceilings; very flexible services distribution (some mechanical ventilation was inevitable); and a precast reinforced concrete frame for quality and speed of erection. As precast concrete has a long lead-in time we made the bottom storey of the building, which contained plant rooms, out of *in situ* concrete so that it could go up while the precast units were being made. As the site was large we precasted complete 7.2 m square pyramids (Figure 6.6) on site, while the rest of the frame (Figure 6.7), including the beautifully finished columns, was cast in a factory.

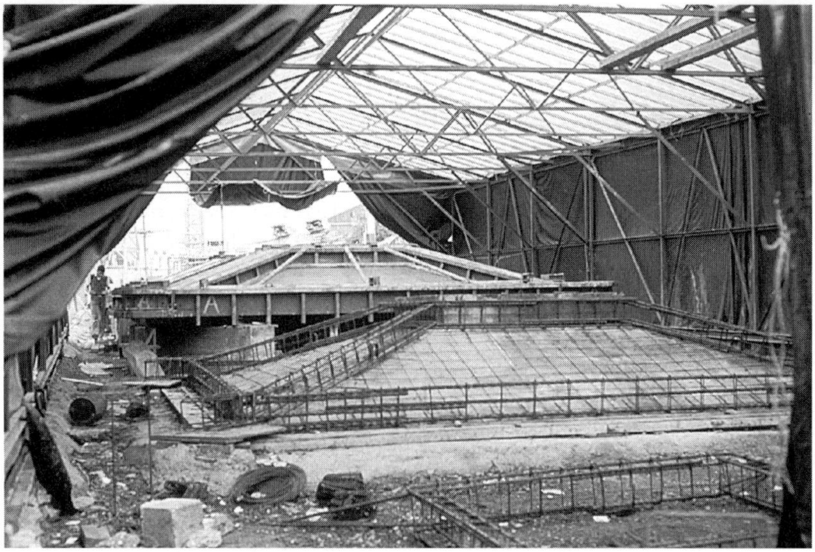

*Figure 6.6 Lloyd's building, Chatham – precasting pyramid units on site.*

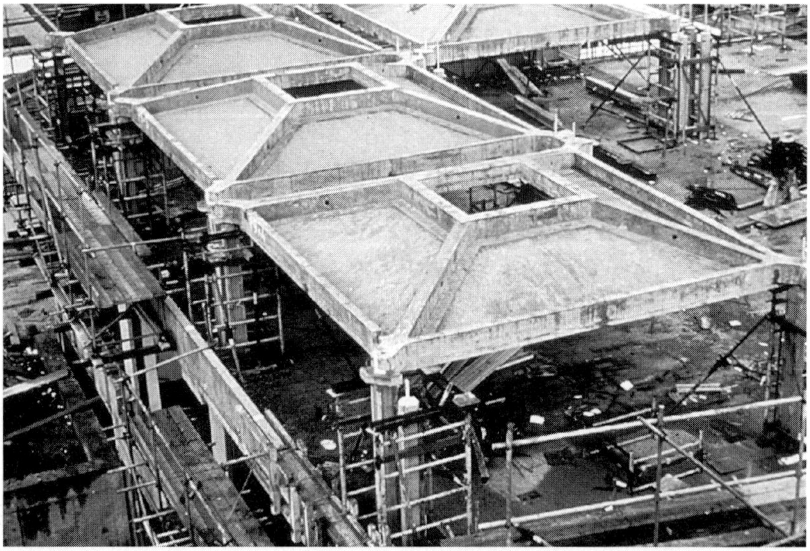

*Figure 6.7 Lloyd's building, Chatham – erection of the precast frame.*

I think that the story of the design and construction of this building really does demonstrate what can be achieved by interdisciplinary collaboration; and the commitment of the group and the belief in what we were doing was impressive. So why, after nine years or so in Arup

Associates, did I want to leave and set up my own practice as a structural engineer?

There are a number of answers and there isn't space here to discuss them all. I will just say now that I felt that I wanted to get away for a bit from big organizations; I wanted a change of scale and a greater variety in the sort of work. I also wanted to try to see what it was like to really earn my own living.

## Price & Myers

So in 1978 Robert Myers and I returned to the world of pure structural engineering and set up Price & Myers. We certainly achieved the aims that I have just listed. The scale changed, the variety increased enormously and we definitely found out about earning our own living – there is something very satisfying about sticking your own stamps on! But what about interdisciplinary collaboration? Well we found, of course, that this varied enormously, like the work, from job to job and design team to design team. In the normal conventional setup of separate designers from different firms coming together to design a building, the framework for communication is almost always set by the architect. The architect is invariably the first one of the designers to be appointed, or at any rate approached, by the client, and the architect usually decides how soon to involve the others. Some architects like to get a design sorted out and agreed with the client before talking to anyone else; some are happy to bring everyone together before anything is decided. We, of course, very much prefer the latter arrangement.

The relationships between the designers do take time to build up. On the first job together everyone is a bit wary of the others, and it is only through developing understanding and finding out how far one can go that a really rewarding relationship can be developed. This applies as much, of course, to the multi-disciplinary group in Arup Associates as to the more usual world of separated practices. In fact, small practices realize that they are vulnerable in a big, competitive world and do feel the need to make good close friendships with others, so that in some ways the incentive for collaboration is quite as strong as in the expected, taken for granted, setup of the multi-disciplinary group. Anyway, in Price & Myers we now have many continuing collaborative relationships with friends in all the other disciplines – including the builders.

### Lord's Cricket School

The building that I have chosen as an example of collaboration between my firm and others is the Cricket School at Lord's (Figures 6.8 and 6.9).

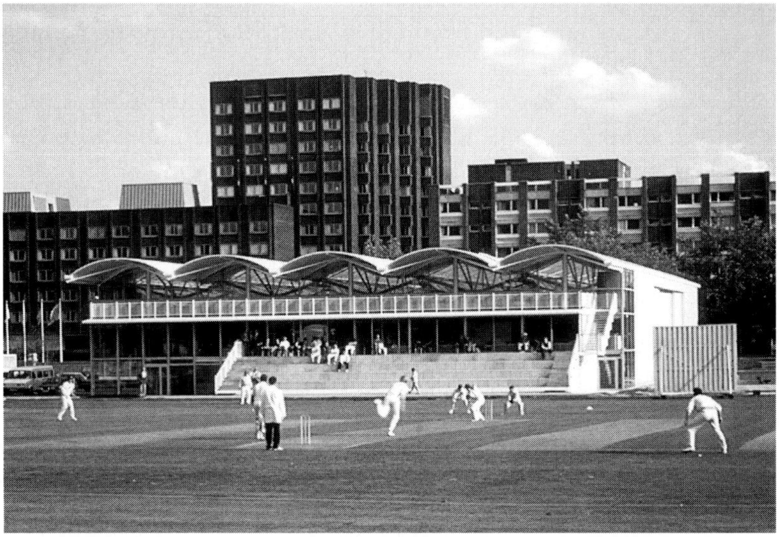

*Figure 6.8 MCC Cricket School at Lord's – view from the practice ground.*

The architect was David Morley, who we had not worked with before, the services engineers were Max Fordham & Partners, who we knew well and had worked with frequently, and the quantity surveyors were Davis Langdon & Everest, who we also knew well.

The crucial question for this building was the lighting. The MCC (owners of Lord's) wanted a high and steady degree of illumination, and this strongly implied artificial lighting within a building with a solid roof, so that the lighting level would be totally independent of the external weather conditions. Right from the start David Morley wanted to use natural lighting if at all possible, as artificial lighting would be very extravagant in energy. Max Fordham said he was convinced that, with the correct orientation of rooflights and suitable reflective screening, very high and steady degree of illumination would be achieved by natural daylight for a large part of the year. The design of the roof – its shape, its covering, the rooflights and the structure – therefore became a matter of crucial importance and the essence of the interdisciplinary collaboration, which explored a number of different ideas in some detail before settling on the scheme that was built. The design that emerged has a clarity and a beauty that come out of fulfilling the purely functional requirements of the lighting. Of particular structural interest is the support of this big roof by spanning the length, end to end, rather than the more obvious width, so that the roof trusses (Figure 6.10) run with the rooflights and form the geometry of the profile of the roof covering.

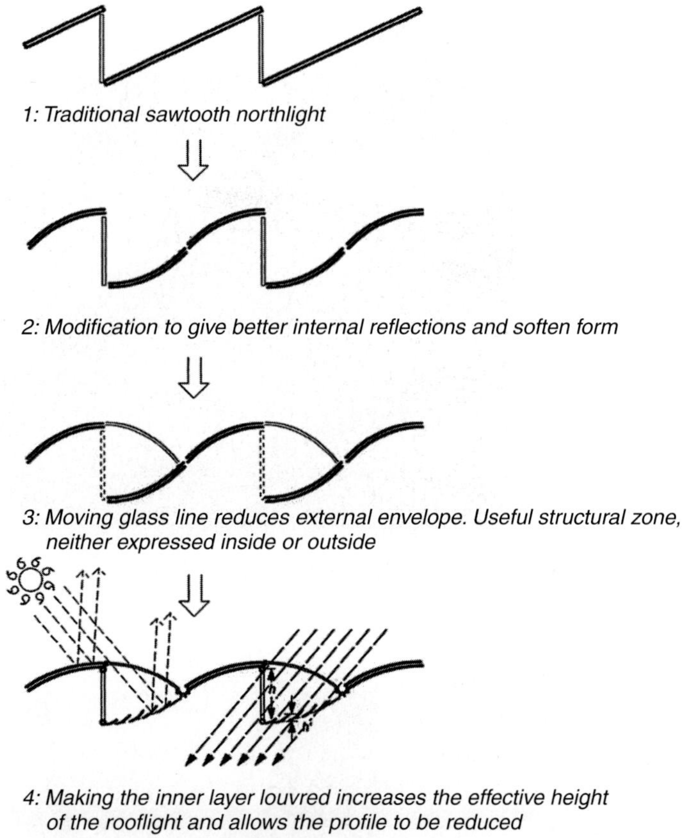


- *Light from CIE overcast sky is 5000 lux*
- *Light flowing through vertical service = 5000/2.5 lux*
- *Light through a vertical rooflight of height h = h × 5000/2.5 lux*
- *In this case light enters through the louvres height h′*
- *The sum of h + h′ is 4.4. The pitch is 7.2 m*

*Therefore the light level is 4.4/7.2 × 5000/2.5 = 1200*

ROOF CONCEPT DEVELOPMENT

*Figure 6.9 MCC Cricket School at Lord's – development of the roof section.*

This rhythm is also related to the width of 'nets' (Figure 6.11) which are suspended off the bottom booms of the trusses.

## Conclusion

I have tried to show how three different organizations of people tackled the issues of interdisciplinary collaboration and how three buildings were produced by their efforts. I am sure that, in the end, there is no one

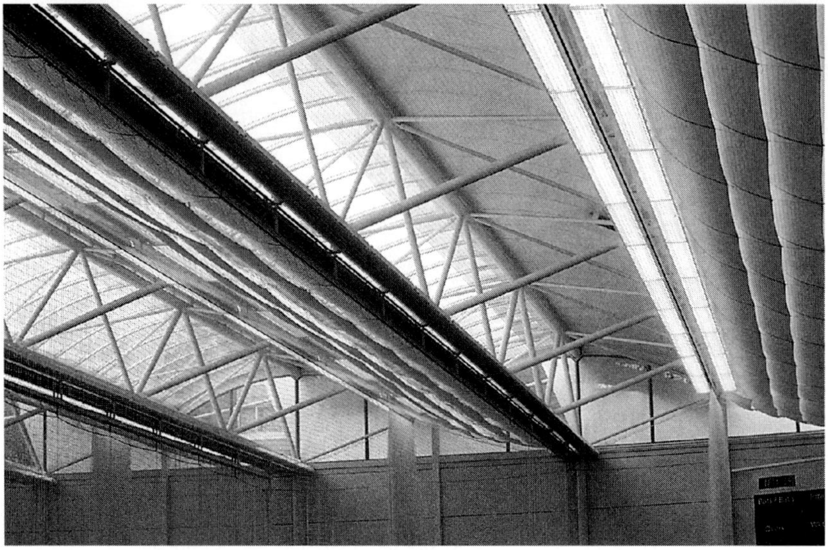

*Figure 6.10 MCC Cricket School at Lord's – roof trusses spanning the length of the building.*

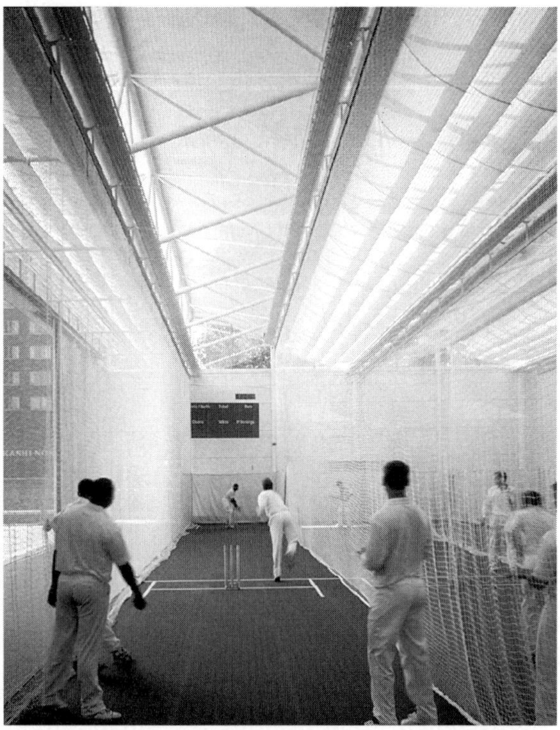

*Figure 6.11 MCC Cricket School at Lord's – interior showing the nets.*

perfect way of going about this. We are all individuals and we all have our own ways of working and communicating with others. What is necessary, regardless of the organization, is the desire to communicate and the passion to build something that is good. The analytical part of the design process may come from the head, but the collaboration and the passion come from the heart.

## Further reading

Royal Fine Art Commission (1995) *Design Education for Engineers*. Thomas Telford, London.

# Part 2
# Educating the interdisciplinary designer

# Introduction

## *Paul Kirby*

Design needs team players with an ability to exploit the benefits of integrated thinking, shared goals and collaborative attitudes. Employers in particular state this need. But how are these attributes to be nurtured? How can a breadth of vision be maintained when early specialization can narrow horizons?

Educators face conflicting pressures. On the one hand is the increasing volume of knowledge germane to each discipline (and needing to be taught); on the other is the need for the breadth of learning that makes communication possible. It is this tension that raises fears of a lack of acuity in the generalist and a lack of integrating skills in the specialist.

Interdisciplinarity is not, however, simply a matter of ensuring that design professionals can work effectively together. Social change, a global market and pressures on the natural environment are some of the forces creating questions of ever greater complexity. These new questions demand a breed of policy-makers and designers able to bring clarity to this complexity and able to find solutions that respect cultural and technical interdependencies.

These are not trivial concerns, and may not, perhaps, be easily remedied by changes to the pattern of professional development at a single stage, for example at undergraduate or postgraduate level. Instead, the industry may need to consider how an interdisciplinary ambition may be kept in sight throughout an individual's professional development.

This part provides a commentary on the current state of professional education and offers examples of recent initiatives in universities and in the workplace, which aim to bring some progress towards interdisciplinary thinking and practice.

In Chapter 7 David Gann and Ammon Salter's detailed analysis of the current state of professional education provides a valuable foundation of data, illustrating present problems and disturbing trends. They present a changing professional environment whose new demands are requiring new knowledge and skills. But where, Gann and Salter ask, are graduates of suitable quality and in suitable numbers to come from? Turning to interdisciplinarity itself, the chapter offers some evidence that universities are moving towards broadening the range of subjects taught; however, Gann and Salter suggest that what is happening at the leading edge and the trailing edge of education may be very different.

Ove Arup was throughout his life a passionate advocate of interdisciplinary attitudes and methods, and in Chapter 8 Sir Jack Zunz describes how the Ove Arup Foundation seeks to keep alive this vision of quality in construction achieved through cooperation. At a different scale, the world faces the twin pressures of rapid population growth and increasing per capita consumption of resources. Sir Jack argues that it is designers, practised in collaborative and inventive decision making, who must take up this challenge. He explains how such considerations led to the Foundation supporting two postgraduate courses, each with a differing thematic structure but both advancing the cause of design in an interdisciplinary environment.

In Chapter 9 Paul Kirby discusses Cambridge University's Master's Course in Interdisciplinary Design for the Built Environment, with which he has been involved since its inception in 1993. The chapter draws on his experience of interdisciplinary education to offer some observations about the nature of interdisciplinarity. One clear message that has emerged is that knowledge of the methods and priorities of other disciplines is not enough to ensure effective collaboration. What is equally important is a set of attitudes including generosity, curiosity and an ambition to work towards the best rather than the adequate.

In the final chapter in this section, John Steele, Andy Parker and Paul Kirby describe how one multi-disciplinary practice, AMEC, recognized that truly collaborative working is not guaranteed by the different disciplines simply sharing office space. 'Seamless Design' is not an inevitability but is the result of the purposeful fostering of cross-disciplinary communication and mutual respect. In consequence AMEC embarked on a programme of in-house training whose purpose was to break down the barriers between disciplines. John Steele, himself an employee of AMEC and a contributor to interdisciplinary research, describes the company's ambition and the training initiative that resulted.

## CHAPTER SEVEN

# Interdisciplinary education for design professionals

*David Gann and Ammon Salter*

### Introduction

Design activities are becoming more complex, creating new requirements for designers to work in interdisciplinary environments. This chapter describes the main reasons for these changes, showing a widening gap between requirements for new design skills and current educational provision in the UK. There is compelling evidence to suggest that design and engineering skills need to be modernized. There is a need for rigour within disciplines, and at the same time interdisciplinary skills will need to be developed progressively throughout people's careers. The education system is itself in a state of flux in response to new forms of knowledge production and the growth of the information economy. Approaches to education and professional training will have to change in order to meet new challenges for design in the built environment in the information age.

The chapter is based on evidence collected in a study of educational requirements of built environment professionals and graduate and postgraduate course provision in British universities. The study was carried out in 1997 and 1998 and some of the information was updated in 2000 (Gann and Salter, 1999). It covers the following main disciplines: civil engineering, architectural design and engineering (including a range of technical specialisms such as structural, mechanical and electrical engineering), building and construction management, surveying and planning.

The chapter begins with a discussion of the new demands upon design professionals, focusing on emerging markets and the need to design different types of buildings and offer new support services. The second part explores changes in the ways in which knowledge is produced and used. This is related directly to the rise of the information economy, which is affecting both the education of new designers and the ways in which practitioners obtain information and ideas required to solve design problems. The third part presents evidence from our survey of education provision. It shows that the British education system for built environment professionals is in a state of crisis, with too few people applying to courses and confusion about the quality and nature of education provision. The chapter ends with recommendations for improving practices, identifying action that needs to be taken by the design professions and education establishments.

## New markets, new skills

The nature and content of design activities in the built environment began to change quite fundamentally in the late 1960s and early 1970s. These changes were associated with the emergence of the information economy and new types of demands placed upon designers to produce buildings and structures to accommodate information intensive activities (Castells, 1989; Duffy, 1997; Gann, 2000; Graham and Marvin, 1996). By the 1980s it had become evident that information and communication technologies (ICTs) were playing a part in reshaping the ways in which people lived and worked. New systems were being installed in many types of buildings, from high-tech offices to retail outlets, doctors' surgeries and homes (Gann *et al.*, 1999). The layout and use of space in many types of building were changing in order to accommodate new ways of working. For example, flexible, lightweight, multi-purpose space was increasingly required and there was growing demand for complex buildings such as silicon chip fabrication plants, biotechnology laboratories, digital control centres, dealer rooms, airports and events stadia. The amount of mechanical, electrical and electronic equipment installed within buildings grew as a proportion of the total value of construction – (from around 7% in the mid-1970s to more than 20% by 1990).

Changes in design and construction processes also took place and international competition for larger projects grew across OECD countries. New forms of project finance emerged, such as the use of private finance to fund what were formerly public buildings. Under these conditions, designers had to consider new issues concerning the long-term operation and use of buildings. Requirements were changing from initial provision of physical buildings to the design of integrated systems in which products and services were bundled together to operate over time within different environments. By the 1990s, designers had to work within a new regulatory framework aimed at reducing the environmental impact of buildings. Issues of 'sustainability' and the use of brown field sites came to the fore and many projects were concerned with work to existing buildings in which designers needed skills and knowledge for refurbishment, modernization and additions to buildings. Their knowledge needed to encompass several vintages of technology and at the same time consider the integration of new systems.

The range of technical solutions expanded and new professional disciplines emerged, focusing on specialist design and engineering tasks such as fire engineering, vertical transportation, acoustics, lighting, building controls and facilities management systems. Managing design processes in which many different types of specialist knowledge had to be integrated was becoming more difficult. ICTs were increasingly being used to assist in the automation of some design tasks and to provide new knowledge about likely performance of design, through the use of simulation and virtual prototyping. However, when used within the context of traditional design practices, these new electronic design tools were as likely to add to the complexity of coordinating design tasks as they were to relieve it.

Significant changes in design and construction processes were promoted, in part through recognition that the traditional separation of design from production did not serve clients and building users well. The traditional approach often meant a poor understanding of user requirements, and weak organization and coordination of design skills, resulting in high costs, delays, wasteful processes and the delivery of buildings which did not perform to specification (Higgin and Jessop, 1966; Latham, 1994; DETR, 1998). Moreover, it was difficult for designers to learn from one project to the next, and mistakes and poor practices were therefore often repeated across many types of building.

In summary, the breadth and depth of skill requirements for built environment designers expanded considerably in the thirty years between 1970 and 2000. New specialisms emerged, but general skills are also needed for coordination, balancing divergent interests and systems integration – see Table 7.1. Many designers require a comprehensive knowledge of previous vintages of technology, systems, structures and buildings constructed in the past. There is a demand for rigour within disciplines and a better understanding of the content of design knowledge across disciplines.

## Specialist and interdisciplinary skills and knowledge

The tendency toward specialization generates its own requirements for new general skills for coordinating and integrating the work of various

*Table 7.1 Examples of new specialist and general skills (source: Gann, 2000, p. 226).*

| New specialist skills | New general and integrative competencies |
|---|---|
| Brief development and definition | Environmental planning |
| Design management | Transport planning |
| Production planning, assembly and installation management | Space planning and changing working patterns |
| Specialist project finance | Business analysis |
| Specialist legal advice | Dynamics and complex systems analysis |
| Risk assessment and management | Building economics and life cycle analysis |
| Safety management | Team building, co-locating, concurrent engineering |
| Supply chain management | Partnering and supply chain management |
| Procurement and logistics | Interdisciplinary skills to integrate engineering and social science expertise |
| Instrumentation and control systems | Understanding users and regulatory frameworks |
| Non-destructive testing | Delivery of integrated products and services |
| Facilities management | |
| Energy management | |
| Water management | |
| Building physics | |
| Materials science | |
| Contaminated land engineering | |
| Geotechnical engineering | |
| Structural engineering | |
| Façade engineering and design | |
| Mechanical and electrical engineering | |
| Heating, ventilation and air-conditioning | |
| Manufacturing engineering | |
| Wind, seismic and vibration engineering | |
| Fire engineering | |
| Lighting design | |
| Acoustical engineering | |
| Simulation and modelling | |
| Computational fluid dynamics | |
| IT systems and data management | |
| Documentation control | |
| Machinery operation and maintenance | |

actors. The skills required for coordination include an understanding of specialist activities and the ability to integrate and manage different sets of knowledge in order to produce successful results. The need for two types of interdisciplinary skill is becoming evident. First, there is a need to understand different scientific and engineering fields. In this sense, interdisciplinarity refers to knowledge across different engineering domains. Second, there is a need to understand knowledge between engineering and social sciences and the humanities. This form of interdisciplinary thinking requires an appreciation of different methods and approaches to problem definition and solving. In both instances, interdisciplinarity involves education which encourages integration of different perspectives into a unified whole. Interdisciplinary work therefore results from the joint and continuously integrated effort to combine two or more specialisms. It involves joint coordinated efforts to integrate lessons from different disciplines (Cooper, 1997).

Interdisciplinary skills therefore complement the emerging pattern of specialization. They provide the means to make connections across disciplines so that processes of design and construction can be seamless. Interdisciplinary skills are also particularly important for problem-solving in areas where there are a large number of variables, together with high levels of uncertainty and risk. As Nobel Laureate Gunner Mydral commented 'problems do not come in disciplines'. Lack of communication across the professions has hindered problem-solving in construction for many years (Higgin and Jessop, 1966). Moreover, new technologies hardly ever function in isolation. Time and again, achieving productivity improvements from a given innovation depends upon the question of availability of *complementary technologies* (Rosenberg, 1982, pp. 58–60). This is even more the case when considering technological systems, such as construction products and processes, where improvements in performance in one part are of limited significance without simultaneous improvements in other parts.

From a business management perspective, firms increasingly require both specialist and generalist skills: they are not an either/or. For example, businesses specialize in order to limit uncertainty in their operating environment. They also attempt to maintain a broader base of more general capabilities in order to manage unknowns when they occur. General managers are needed to balance these two requirements.

Developing these new specialist and general, interdisciplinary skills has not been easy in the UK. Traditional customs and practices, and legal and institutional frameworks have hindered the emergence of new approaches. In many ways, professional institutions have acted as the protectors of traditional skills, and have maintained standards through professional training and education programmes. Their focus has usually been on specialist knowledge pertaining to particular traditional disciplines. The benefits of this specialization have been tremendous. Professional institutions have raised standards, supervised new developments in the sector and institutionalized quality control. Yet, such specialization has also led to a breakdown of understanding between different actors involved in design and construction processes. As each field has become more specialized, they have developed their own technical language and understanding. They have also guarded entry against what they perceive to be 'external' interests. Demarcation lines have therefore been maintained. This has hindered communication and the transfer of knowledge across disciplinary boundaries.

## A new system of knowledge production

Recent growth in specialist design, engineering and construction skills has occurred during a period of broad changes in the ways in which knowledge is produced and used. This is associated with the evolution of the information age and with changing working patterns such as the likelihood that people will work for a number of different employers, rather than one, and that they may often have several careers throughout their working lives. There has been a shift from the more rigid and certain employment patterns of the previous 'corporatist' machine age towards insecurity, uncertainty and in some ways a more exciting mode of work in the information age (Castells, 1996; Sennett, 1998). These broad, sweeping social, economic and technological changes have affected the conditions within which engineers, designers and managers work. They have also brought about major changes in the ways in which they are educated and trained.

Gibbons *et al.* (1994) argue that a new system of knowledge production has emerged, and this has radical consequences for the ways in which people learn and are taught, carry out research and solve problems. The role of traditional institutions such as universities and professional associations is brought into question by different ways of producing, sharing and using knowledge using ICTs. The key features of the shift from Mode 1 to Mode 2 are shown in Table 7.2.

Gibbons *et al.* argue that the development of Mode 2 places new pressures on universities to adapt their educational provision to fit new patterns of knowledge production. The breakdown of traditional disciplines and overlapping nature of technological change will force universities to develop new integrated course offerings that reflect the practices of the industry. These new integrated course packages will be trans-disciplinary by operating above traditional disciplines and involve considerable problem-oriented training. Architects, building services engineers, surveyors and other built environment professionals will be grouped together in integrated educational courses, focused around projects. This new model would involve considerable

*Table 7.2 Modes of knowledge production (source: adapted from Gibbons* et al., *1994).*

| Mode 1 | Mode 2 |
|---|---|
| • Discipline-based teaching<br>• Clear demarcation between universities and industry, academic and consultant<br>• Universities educated, industry trained<br>• More students means a better education system<br>• Peer review<br>• High levels of trust in science – independence | • Interdisciplinary learning<br>• Blurring boundaries between universities and industry, academic and consultants<br>• Greater collaboration<br>• Knowledge production wide-spread in society<br>• Learning organizations<br>• Research in the context of application<br>• Declining trust in science and scientists |

change in the current system of discipline-based education provision. It will force universities to reconsider traditional demarcations between built environment professionals.

A recent study by the Institute of Employment Studies, commissioned by the EPSRC and the Institute of Materials, surveyed employers' views about the provision of newly qualified postgraduate engineers and material scientists and highlights the need for new skills in engineering (Jagger and Conner, 1998). The study involved 50 face-to-face interviews with engineering employers and a survey of 1600 recent graduates from masters and doctoral engineering and materials science training programmes. It showed that employers look for particular 'soft skills' among graduates; these include:

- interpersonal skills
- communication skills
- business awareness
- flexibility/versatility
- team working
- initiative/proactivity.

In terms of technical skills, employers felt that these skills were important but not of overarching concern: current graduates in materials science were found to have adequate levels of technical competence. The ability of graduates to translate technical knowledge into non-technical language was considered a prize asset among employers. In terms of the construction sector, the employers interviewed stressed the importance of soft skills in combination with technical competence (Jagger and Connor, 1998).

This study also showed that the most important skills students gained in relation to their current activities were learnt whilst studying for Masters or Doctor of Philosophy degrees. The results are based on respondents assessing the contribution of a list of skill categories on a 1 to 7 point Likert scale. On the basis of the responses, the skills currently most important for graduates appear to be problem-solving, communication skills and business awareness. For Ph.D. students, the most important skills learnt were in communication, IT skills and problem-solving (see Table 7.3). The patterns for both Masters and Ph.D. graduates were similar. What is interesting about this finding, for our purposes, is that it shows the skills most important for graduates in engineering tend to be interdisciplinary, associated with activities outside narrow technical disciplines (Jagger and Connor, 1998).

As part of the IES study, respondents were asked to describe what skills they had gained as a result of their education. Overall, the fields that graduates felt were important in their current practices were those

*Table 7.3 Skills currently important for Masters and Ph.D. graduates (source: IES survey of recent engineering and materials science postgraduates).*

| Skills | Masters | Ph.D. | All |
|---|---|---|---|
| Communication skills | 6.0 | 6.1 | 6.0 |
| Problem-solving | 6.1 | 6.0 | 6.0 |
| Planning and organization skills | 5.8 | 5.9 | 5.9 |
| Oral communication | 5.9 | 6.1 | 5.9 |
| Problem identification | 5.7 | 5.9 | 5.8 |
| IT skills | 5.6 | 6.0 | 5.7 |
| Initiative/proactivity | 5.7 | 5.8 | 5.7 |
| Time management skills | 5.7 | 5.8 | 5.7 |
| Teamworking skills | 5.7 | 5.3 | 5.6 |
| Flexibility | 5.6 | 5.3 | 5.5 |
| Writing skills | 5.3 | 6.0 | 5.4 |
| Project management skills | 5.4 | 5.5 | 5.4 |
| Business awareness | 4.8 | 4.9 | 4.8 |
| Creativity | 4.6 | 5.0 | 4.7 |
| Customer orientation | 4.8 | 4.6 | 4.7 |
| Leadership | 4.5 | 4.7 | 4.6 |
| Literature searching and review | 4.0 | 4.6 | 4.1 |
| Use of test/analytic equipment | 3.9 | 3.6 | 3.8 |

*Table 7.4 Skills gained from education on Masters and Ph.D. programmes (source: IES survey of recent engineering and materials science postgraduates).*

| Skills | Masters | Ph.D. | All |
|---|---|---|---|
| Problem-solving | 5.1 | 5.6 | 5.2 |
| Problem identification | 4.8 | 5.4 | 4.9 |
| Initiative/proactivity | 4.8 | 5.5 | 4.9 |
| Literature searching and review | 4.6 | 5.7 | 4.9 |
| Communication skills | 4.8 | 5.1 | 4.8 |
| Planning and organization skills | 4.7 | 5.1 | 4.8 |
| IT skills | 4.6 | 5.5 | 4.8 |
| Oral communication | 4.7 | 4.9 | 4.7 |
| Writing skills | 4.4 | 5.6 | 4.6 |
| Time management skills | 4.4 | 4.5 | 4.4 |
| Project management skills | 4.3 | 4.2 | 4.3 |
| Teamworking skills | 4.4 | 3.2 | 4.2 |
| Creativity | 4.0 | 4.8 | 4.2 |
| Flexibility | 4.0 | 4.6 | 4.1 |
| Use of test/analytic equipment | 3.9 | 4.2 | 4.0 |
| Leadership | 3.6 | 3.2 | 3.5 |
| Business awareness | 3.2 | 2.9 | 3.2 |
| Customer orientation | 2.6 | 2.6 | 2.6 |

developed in their formal education. Graduates gained problem-solving skills, initiative, literature searching and communication skills from their education (see Table 7.4). The results show that, on the whole, the education system for engineering and materials science is meeting the needs of employers and recent graduates. But there are significant gaps. For Masters graduates, skills in customer orientation, leadership and business awareness seem to be not so well developed and yet important to their current work practices. For doctoral graduates, customer orientation, business awareness, flexibility and leadership seem to be poorly developed and yet are important for their current employment practices (Jagger and Conner, 1998). This suggests areas for improvement within the education system.

## Where we stand in the UK

Dual pressures are at work, changing the knowledge and skills of built environment designers, engineers and managers: those emanating from practice and the increase in specialization, and those caused by changes in the ways in which knowledge is produced and used more generally. The use of information and communication technologies often underpins these changes. In this section, we review the current status of the built environment education system in the UK. First, we explore the general patterns of applications, acceptances and graduates. Second, we investigate the extent of interdisciplinary training in current course offerings, suggesting that education for interdisciplinary skills has already become part of the strongest courses in the UK.

The built environment education system has been in a general crisis since the mid-1990s. In 1997, there were 457 undergraduate courses at 63 universities in the UK offering courses in the built environment. There was considerable overlap, under-staffing, and fierce competition among these programmes. By 2000, many departments were closing.

The crisis has been fuelled by a steep decline in applications to courses in the built environment, in particular civil engineering and building and construction. From 1994 to 1999, applications in civil engineering fell by 40%. Building and construction courses were down 41% and architecture was down 14%. This fall in applications occurred at a time when there was a general increase in the total number of applications to the UK university system (Table 7.5).

While the total number of applications has fallen, the number of acceptances to courses in the built environment has also declined. From 1994 to 1999, civil engineering acceptances were down 24% and building and construction fell by 7%. However, architecture increased its number of acceptances by 18%. Again the built environment subject groups were only able to accept fewer students than the average of the entire UK education systems (Table 7.6).

What has been the impact of this decline in applications and acceptances? One way to gauge the impact of these changes is to compare the number of acceptances with the number of applications. This provides a

*Table 7.5 Number of applications in built environment courses (source: UCAS, 1999).*

| | 1994 | 1995 | 1996 | 1997 | 1998 | 1999 | Change |
|---|---|---|---|---|---|---|---|
| Civil engineering | 5 104 | 4 538 | 4 207 | 3 766 | 3 442 | 3 080 | –40% |
| Architecture | 3 269 | 3 237 | 3 097 | 3 179 | 3 004 | 2 797 | –14% |
| Building/ construction | 4 006 | 3 860 | 3 232 | 2 803 | 2 671 | 2 369 | –41% |
| Built environment | 12 379 | 11 635 | 10 536 | 9 748 | 9 117 | 8 246 | –33% |
| All subject groups | 405 117 | 419 442 | 418 400 | 458 781 | 446 457 | 442 931 | 9% |

*Table 7.6 Number of acceptances into built environment courses (source: UCAS, 1999).*

| | 1994 | 1995 | 1996 | 1997 | 1998 | 1999 | Change |
|---|---|---|---|---|---|---|---|
| Civil engineering | 3 453 | 3 347 | 3 218 | 3 143 | 2 869 | 2 624 | –24% |
| Architecture | 1 972 | 2 159 | 2 174 | 2 368 | 2 311 | 2 333 | 18% |
| Building/ construction | 2 367 | 2 718 | 2 430 | 2 498 | 2 345 | 2 211 | –7% |
| Total built environment | 7 792 | 8 224 | 7 822 | 8 009 | 7 525 | 7 168 | –8% |
| Total all subject groups | 247 567 | 264 738 | 273 032 | 308 236 | 302 683 | 307 677 | 24% |

rough measure of quality, indicating the demand for various courses. What becomes apparent when the built environment is subjected to this form of analysis is that there has been a significant shift in the character of students on the courses. From 1994 to 1999, the percentage of acceptances by applicants in civil engineering rose from 65% to 85% (Table 7.7). This pattern was mirrored in architecture (60% to 80%) and building and construction (59% to 93%). This pattern contrasts with aggregate data compiled for all courses in the UK where the shift in the pattern of acceptances by applications changes slightly from 61% to 69% over the same period of time. This suggests that quality of entrants into courses in the built environment is likely to be declining.

The reason for the decline in applications in the built environment is varied. But one of the strongest factors is the image of the industry itself. It is often seen as dirty, dangerous, white and male. This is reflected in the percentage of women and ethnic students in built environment professions. The perception is that entry into construction is a more

*Table 7.7 Percentages of acceptances by applications for built environment courses (source: UCAS, 1999).*

| | 1994 | 1995 | 1996 | 1997 | 1998 | 1999 |
|---|---|---|---|---|---|---|
| Civil engineering | 68 | 74 | 76 | 83 | 83 | 85 |
| Architecture | 60 | 67 | 70 | 74 | 77 | 83 |
| Building/construction | 59 | 70 | 75 | 89 | 88 | 93 |
| Total built environment | 63 | 71 | 74 | 82 | 83 | 87 |
| Total all subject groups | 61 | 63 | 65 | 67 | 68 | 69 |
| Difference: Total built environment and all subjects | –2 | –8 | –9 | –15 | –15 | –18 |

*Table 7.8 Percentage of women of total applications in built environment courses (source: UCAS, 1999).*

| | 1994 | 1995 | 1996 | 1997 | 1998 | 1999 |
|---|---|---|---|---|---|---|
| Civil engineering | 12 | 13 | 14 | 14 | 15 | 13 |
| Architecture | 24 | 28 | 30 | 32 | 35 | 35 |
| Building/construction | 9 | 11 | 11 | 11 | 12 | 12 |
| Total built environment | 14 | 16 | 17 | 19 | 21 | 20 |
| Total all subject groups | 50 | 51 | 51 | 52 | 52 | 53 |

risky choice than entry into professions such as medicine, law, accountancy or even general engineering. A key factor in explaining these patterns of decline is the limited role of women in the professions. From 1994 to 1999, the percentage of women applicants in civil engineering and building and construction barely changed. The percentage of women applicants in architecture increased from 24% to 35%. This partly explains the better performance of architecture in attracting students. However, all these subject areas compare poorly to the overall percentage of women in all other subjects. From 1994 to 1999, the percentage of women applicants to the UK university sector rose from 50% to 53% (Table 7.8). The lack of women applicants is reflected in the fact that women account for less than 1% of members of professional institutions.

The situation is complicated by the difficulty that potential built environment students experience in choosing which course of study to pursue at which university. The UCAS handbook (1998) lists a huge number of courses under a large number of categories at numerous universities. This presents a major difficulty in that it is currently not easy to assimilate what courses are on offer across the board. Neither is it easy to assess which degrees might offer a more forward-looking approach with interdisciplinary courses designed to meet the changing needs of industry. By itself, the presentation of data on courses in the UCAS handbook is likely to be confusing to potential students wishing to make informed choices. The likelihood that a potential student can identify and consider seriously the opportunities in relatively new and exciting areas such as building services must be regarded as remote. There has also been extreme confusion about SARTOR (Standards and Routes to Registration), the difference between incorporated and chartered engineering courses, and the myriad of professional institutions each with its own rules and procedures.

Another key feature of the UK built environment education system in the late 1990s is the role of overseas students. From 1994 to 1999, the

*Table 7.9 Percentage of foreign students in UK built environment courses (source: UCAS, 1999).*

| | 1994 | 1995 | 1996 | 1997 | 1998 | 1999 |
|---|---|---|---|---|---|---|
| Civil engineering | 20 | 27 | 33 | 39 | 37 | 35 |
| Architecture | 19 | 25 | 28 | 30 | 30 | 31 |
| Building/construction | 10 | 14 | 14 | 13 | 11 | 14 |
| Total built environment | 17 | 22 | 26 | 28 | 27 | 27 |
| Total all subject groups | 8 | 9 | 10 | 10 | 10 | 12 |

percentage of acceptances of overseas students increased in civil engineering from 20% to 35%. This pattern was mirrored in architecture (19% to 30%). There was little change in building and construction, from 10% to 14% (Table 7.9). Although the participation of overseas students obviously provides many benefits to the education system, the question remains as to what percentage of overseas students remain working in the UK after finishing this education and therefore contribute to improving the performance of the UK construction sector.

As part of our study, we developed a database designed to assess the extent of interdisciplinarity in courses offered in each educational programme. The research drew from a listing of classes offered in each course each year, as described in the UCAS handbook. Interdisciplinarity was defined as the number of classes offered each year to students outside their core disciplinary training. For each year of each course, a two-point scale was developed for the extent of interdisciplinarity. The two-point scale represented the extent of interdisciplinary course work students were asked to complete for that particular year. Each year was assessed and the total was taken for each year of the course. The total was then compared with the maximum possible score for each three-, four- or five-year programme. This figure represents the *interdisciplinary index* for a particular course. For example, if a four-year civil engineering course offered two courses in construction management in the fourth year, it would receive a two-point score for the fourth year, and a two-point score out of eight for the total years of the course. Having developed an interdisciplinary index score for each course, it was possible to see what features characterize interdisciplinary courses.

It is necessary to raise a note of caution about the limitations of such a method of calculating interdisciplinarity. First, it says little about the actual content of each class offered in the course. Second, it misrepresents interdisciplinary courses, such as construction management and to some extent architecture, because it focuses on courses offered outside the discipline. If the course is itself interdisciplinary then it does not reflect this fact. Despite these limitations, it does offer a relatively comprehensive measure of interdisciplinarity in UK built environment courses because it is based on classes offered each year in each course. It also enables the comparison of interdisciplinarity with other factors of university behaviour.

The first finding when analyzing courses using this interdisciplinary index is that interdisciplinarity is widespread in UK built environment courses. The average student receives at least one class outside his or her subject area per year on their course. As might be expected, construction management at undergraduate level has the highest level of interdisciplinarity among the five course groups (building services/ built environment, construction management, surveying, architecture, and civil engineering) (see Table 7.10).

In general terms, interdisciplinary courses were associated with high entry requirements. Table 7.11 shows that courses with high entry requirements (defined using a three-point scale of the point totals of A levels that students needed to have to qualify for the programme) have greater levels of interdisciplinarity. This finding held in different subject areas, such as civil engineering.

Interdisciplinarity was also generally associated with high scores in the 1996 Research Assessment Exercise. Departments receiving 5 or 5* in the 1996 RAE had a higher level of interdisciplinarity than lowly rated research departments (Table 7.12). This result varied among subject areas, however. Surveying, civil engineering and architecture all exhibited high levels of interdisciplinarity and high RAE scores, yet in construction management and mechanical and electrical systems the opposite was true.

*Table 7.10 Interdisciplinarity in built environment courses for undergraduates, 1997–98 (source: SPRU Built Environment Course Database).*

| Courses | Interdisciplinary index | Number |
|---|---|---|
| Civil engineering | 80% | 145 |
| Architecture | 49% | 78 |
| Building services | 77% | 66 |
| Surveying | 76% | 34 |
| Construction management | 85% | 32 |
| All courses | 72% | 355 |

*Table 7.11 Entry requirements and interdisciplinarity (source: SPRU Built Environment Course Database).*

| Entry requirements | Interdisciplinary index | Number of courses |
|---|---|---|
| High | 79% | 112 |
| Medium | 70% | 200 |
| Low | 68% | 44 |
| All | 72% | 356 |

*Table 7.12 Research assessment exercise scores (1996) and interdisciplinarity (source: SPRU Built Environment Course Database).*

| RAE scores | Interdisciplinary index | Number of courses |
|---|---|---|
| 5 and 5* | 74% | 67 |
| 4 | 76% | 58 |
| 3a and 3b | 72% | 189 |
| 1 and 2 | 66% | 59 |
| All departments | 71% | 373 |

These findings are preliminary; much more detailed research and analysis are required to understand the dynamics between education provision and industrial needs. They indicate that research-intensive departments allow students to draw from more disciplines in their course modules. Research-intensive departments appear more sensitive to interdisciplinarity. In sum, they show that although interdisciplinarity is not a panacea for the needs of the construction sector, it does seem to be an important characteristic of the best courses and departments in the UK built environment education system.

Yet the opposite may not be true. In weak departments or courses, interdisciplinarity may be an unnecessary luxury. Several interviewees argued that the quality of technical skills among many graduates has fallen in many departments. This deterioration, they felt, was due, in part, to an overemphasis on interdisciplinarity in some courses. Technical skills had been displaced by 'fashionable skills', argued one interviewee. Given the falling number of applicants and concerns over the skills of current students, there may be some validity to this argument. It would seem that in the 'tail end' of the education system interdisciplinarity is not essential, or rather it might interfere with necessary technical education. At the other end of the educational system, interdisciplinarity appears to be a welcome complement to a strong technical education.

## Conclusions

In this chapter we have argued that there have been significant recent changes in demands upon designers, engineers and managers responsible for the production and use of the built environment. There has been an increase in specialization and an associated requirement for more generalist knowledge. At the same time, information and communication technologies have in part been responsible for changes in working patterns and methods of problem-solving. Meanwhile, broad, sweeping changes in the ways in which knowledge is produced and used have become evident in many parts of the economy and society.

In consequence, the education system for built environment professionals needs major reform if it is to produce people with the types of skills and knowledge needed by a dynamic, modern industry. Falling numbers and declining standards of applicants to many courses have resulted in the need to teach basic 'refresher' courses. In some cases this has meant teaching an additional year of preparatory studies before the main degree course begins. At the same time there are demands from industry to expand the breadth of teaching material. Many university departments have attempted to squeeze extra interdisciplinary subjects into what are often already crowded course programmes. There are limits to the extent to which interdisciplinary courses can be fitted into traditional course programmes, particularly at the undergraduate level.

An alternative option is to develop new ways of teaching and working that include the development of interdisciplinary teamwork focused on practical projects. Several university departments are moving in this direction, integrating evidence-based courses, applying the concept of forensic engineering and attempting to keep the interest and motivation of high-calibre students.

Yet the decline in applications will not be overcome by simply changing the structure of courses. The roots of the problem lie at the heart of the way the industry currently works in the UK. A few innovative UK design, engineering and construction firms are developing new techniques and practices. They provide an inspiration to new students and professionals. However, there are many laggards and even more firms sitting in the middle position, seemingly incapable or uninterested in changing their practices. The lack of investment in research and development means that most firms lack capacity to capture and exploit new ideas. They fail to attract and retain talented people.

The built environment professions have the potential to innovate and offer creative, rewarding and satisfying careers. One step to overcoming the mismatch between the current state of the education sector and the industry, and the possibility of new technologies to transform how buildings are designed, will be to redefine the roles of professionals. Traditional demarcation lines and the possessiveness of titles and territory will be seen as increasingly outmoded as a new model of the integrated professional emerges.

What is needed is a new approach to developing these skills, involving educationalists, industry and professional institutions. New courses that combine disciplines and the possibility of accreditation across professional institutions would be a signal to new students of the integrated nature of work in the built environment. Industry employers will need to change their recruitment and retention strategies. The traditional mechanisms of compensation linked to professional membership and movement up engineering grades will also need to be reviewed: the rise of new specialisms makes this model of support and promotion outmoded. New, more flexible systems of remuneration and promotion would perhaps attract young people and encourage them to stay within the sector, rather than choose a career in what appear to be more dynamic and rewarding areas such as in multimedia, software, financial services or the life sciences.

There needs to be a serious effort to streamline and amalgamate the current structure of professional institutions in the UK. A more international approach is needed. It is possible to imagine having one professional institution for the built environment, with sub-groups focusing on specialist knowledge domains. This would involve a reassessment of the current system and willingness of professional associations to work more closely together in course development, professional accreditation and continuing professional education.

## Acknowledgements

We are grateful to the Ove Arup Foundation for their sponsorship of the study of interdisciplinary design skills and education. We are also grateful to the EPSRCs who provided support through grant GR/L79465 – Mapping, Measuring and Managing Technology in Construction Firms – and through the IMI/RAEng Chair in Innovative Manufacturing. This funding enabled us to carry out and update the research and to write this chapter.

## References

Castells, M. (1989) *The Informational City*. Blackwell, Oxford.

Castells, M. (1996) *The rise of the network society Vol.1 – The Information Age: Economy, society and culture*. Blackwell, Oxford.

Cooper, I. (1997) Support for interdisciplinary research. *Workshop on Interdisciplinary Research*. University of Salford Research and Graduate College.

DETR (1998) *Rethinking Construction: the Report of the Construction Task Force*, July 1998.

Duffy, F. (1997) *The New Office*. Conran Octopus, London.

Gann, D. M. (2000) *Building Innovation – Complex Constructs in a Changing World*. Thomas Telford, London.

Gann, D. M. and Salter, A. (1999) *Interdisciplinary Skills for Built Environment Professionals*. The Ove Arup Foundation, London.

Gann, D. M., Barlow, J. and Venables, T. (1999) *Digital Futures – Making Homes Smarter*. Joseph Rowntree Foundation/Chartered Institute of Housing, Coventry.

Gibbons, M., Limoges, C., Nowotny, H., Schwartzmann, S., Scott, P. and Trow, M. (1994) *The New Production of Knowledge – the Dynamics of Science and Research in Contemporary Society*. Sage, London.

Graham, S. and Marvin, S. (1996) *Telecommunications and the City: Electronic Spaces, Urban Places*. Routledge, London.

Higgin, G. and Jessop, N. (1966) *Interdependence and Uncertainty*. The Tavistock Institute, London.

Jagger, N. and Connor, H. (1998) *Employers' Views on the Provision of Postgraduate Engineers*. Institute of Employment Studies, Brighton.

Latham, M. (1994) *Constructing the Team. Final Report of the Government/Industry Review of Procurement and Contractual Arrangements in the UK Construction Industry*. HMSO, July 1994.

Rosenberg, N. (1982) *Inside the Black Box*. Cambridge University Press, Cambridge.

Sennett, R. (1998) *The Corrosion of Character – the Personal Consequences of Work in the New Capitalism*. WW Norton and Company, New York.

**CHAPTER EIGHT**

# Design for the built environment: bridging the Two Cultures

*Sir Jack Zunz*

*Wisdom's not a thing you find.*
*It is more like a state of mind*
*And those who have it do not show it*
*While those who show it do not know it-*
*Indeed a most elusive thing!*
*To get what's needed of the stuff*
*A life span isn't long enough.*

This is one of Ove Arup's little verses. He was very accomplished in making short, pithy rhymes. He used this medium to express tersely, and with considerable wit, some profound thoughts – comments on lifelong passions as well as on more topical matters.

This particular one alludes to his passionate search for knowledge and truth. He knew he could never find enough of the former, while his study of philosophy taught him that the latter, though elusive and probably unattainable, was nevertheless worth searching for.

This quest for knowledge and truth was most likely the key to his aspiration towards holistic solutions to problems of design – be they buildings or bridges or chess sets or business arrangements or partnership agreements – the whole fabric of life was to him a question of design. The problems associated with integrating all the emerging and increasingly complex specialized knowledge into a harmonious, human, functional and where appropriate, visual whole, became his lifelong obsession. But he was also very critical, and holistic solutions in themselves were not enough – the quality of everything, tangible and intangible, came under careful scrutiny.

The first half of the twentieth century, which encompassed Ove's formative years, was dominated by the legacy of the nineteenth century and its Industrial Revolution. The explosion of knowledge and expertise led to the erosion of the holistic designer and to the advent of the specialist. Unidisciplinary attitudes evolved and tended towards dominating solutions to interdisciplinary problems. One of Ove's most significant legacies is his passion for design as a total process – he stimulated an interdisciplinary culture where the quality of the whole as well as that of the part became the major determinant.

When Ove died it was thought appropriate to honour his memory, and where possible further some of his ideals, by setting up a Foundation in his name. The Trust Deed states *inter alia* that the objectives of the Foundation are 'for the advancement of education directed towards the promotion, furtherance and dissemination of matters associated with the built environment, with some emphasis on the multi-disciplinary nature of design in engineering and architecture'.

To begin with, the trustees of the Foundation perceived the need for an initiative to explore ways whereby professionals working in the built environment could be brought together to study common objectives under laboratory conditions, where a better understanding of the whole and not simply the part could be coupled with the possibility of widening the knowledge base of the participants. It was thought that this would create opportunities for bright young professionals to work together in a creative environment as well as to enrich their education by experiential contacts with leading academics as well as practitioners from our industry.

It was intended that the learning experience should not be limited to their particular disciplines but be extended to the humanities and other contextual studies. It was felt not only that potential leaders in design and construction would benefit from such a programme, but also that the cadre from which key posts in government and industry related to the built environment are filled would be enhanced. How often do we hear the complaint that these jobs are occupied by lawyers, accountants and historians? Usually this is for the very good reason that the relevant professionals are not thought to be suitable, perhaps too narrow in outlook – the result of their education and training experiences.

These were no more than vague perceptions held by the trustees and their advisors. To explore them and related matters and, if possible, to articulate them more effectively, a seminar was sponsored at Madingley Hall in Cambridge in 1991. To make this two-day seminar interactive for the participants it had, unfortunately, to be limited to about 30 people. It was nonetheless attended by many luminaries from the construction industry.

It is interesting to note that it was thought to be beneficial and enlightening for the participants in the seminar if one or two 'outsiders' were invited – 'outsiders' in the sense that they are not professionals working directly in our industry but who were thought could, because of their particular standing and wisdom, contribute to the debate. One of these was Corelli Barnett, the distinguished military historian. He has written incisively about Britain's industrial decline and its causes. In a post-seminar paper he made some succinct and pertinent comments. He said *inter alia:*

> *The written papers seemed to me somewhat waffly and full of long jargon words, whereas the basic meat could have been expressed with far greater economy and clarity. What does this say about past and present professional education for the built environment?*

This was before Latham and Egan underlined the need for the industry to communicate effectively, for its members to stop fighting each other and to place their clients' needs at the top of the agenda. Barnett's words were a timely homily to be more outward looking and communicate more effectively with the public at large.

The outcome of the seminar was many-sided, but it confirmed that the perception for the need to take some new initiatives in postgraduate education was well founded. The question was where to start. There were obvious financial constraints, so that the first of these initiatives, like charity, began at home. A postgraduate programme open to all the specialist professions in the construction industry was proposed. It could be full-time or part-time and be studio-based.

A number of higher education establishments were asked to express their interest. Nearly all professed enthusiasm for the concept. Six were then asked to provide detailed proposals on the basis of a brief which, while not prescriptive, included some discrete objectives. There was reference to the development of knowledge as well as the self-assurance of the participants. It was to be open to Built Environment Professionals – Architects, Engineers (Civil, Structural, Transportation and Building Services), Quantity Surveyors, Developers and Clients generally, as well as Constructors whatever their basic training and qualifications. It was suggested that the programme should include contextual studies covering a wide range of possible subjects from philosophy, literature and history to the environment, economics and communications. After an exhaustive selection process, it was agreed to support the proposal from the University of Cambridge. The Interdisciplinary Design for the Built Environment (IDBE) Programme is the result.

The sponsors of the Cambridge programme had two strategic objectives. First the basis of the programme should be studio-based and project-oriented. The studio base is essential to bring out the necessary creativity. Projects should be wide ranging with an emphasis on the 'big' issues of the day – urban design, transportation, the environment and sustainability – while some of the more modest interdisciplinary matters in construction should not be overlooked. The second strategic objective was to give the students the opportunity to widen their knowledge base by opening windows to subjects ranging from literature to history, from philosophy to the social sciences.

Confucius said that 'the people may be made to follow a course of action, but they may not be made to understand it'. Many students taking part in programmes for the professions in the built environment follow what they perceive to be a course in design without really understanding it. Sir Alan Muir Wood explained the meaning of design some time ago. It is a singularly helpful explanation of a concept which has such a broad band of interpretations. Here I can do no more than quote its significant beginning: 'Design is the central element of architecture – and engineering. Design should therefore form a central feature of education and training...'.

The most basic definition of the word design appears to be 'a mental plan, an intention'. If design conveys an idea, an intention, then each part of the process, not only the product, is included in the concept of design. The concept of design here, therefore, is the whole, the totality, not just the part, and as Alan Muir Wood said, architects think it is all about style, while engineers think it is all about analysis.

In this context it is also worthwhile recalling C. P. Snow's thesis, articulated in his 1959 Rede lecture about 'The Two Cultures'. He said *inter alia:*

> *In our society we have lost even the pretence of a common culture. Persons educated with the greatest intensity we know, can no longer communicate with each other on the plane of their major intellectual concern. This is serious for our creative, intellectual and, above all, our normal life. It is leading us to interpret the past wrongly, to misjudge the present, and to deny our hopes for the future. It is making it difficult or impossible for us to take action.*

Snow's arguments are not flawless, but there is still too much reality in his proposition for comfort. Snow is credited with coining the phrase 'The Two Cultures'. He spoke eloquently, if controversially, about the matter and its perceived consequences. It can of course be something of an exaggerated generalization. But his proposition is deeply rooted in nineteenth century concepts of educational standards. Some leading academics of the day were not sympathetic to 'useful' education.

It is sadly true not only that professionals working in the built environment often come from opposing ends of these cultures, but also that our clients and the public we serve are no less polarized. The roots of this dilemma are deeply embedded in this country's social and political history. What is unforgivable is that even today it is still aided and abetted by a secondary education system which continues to create a kind of intellectual apartheid in our young school-leavers.

At a critical stage in our civilization when our environment, natural and built, coupled with the enormously complex issue of sustainability and all that this implies, is high on the public agenda, it is downright irresponsible not to do something about the education of our young to bridge these cultures securely, to merge them where possible. It is only by deploying the might of our total intellectual resources, pooling the knowledge and experience of many disciplines, that we can even begin to offer solutions to today's – let alone tomorrow's – problems.

The aim of the Cambridge programme is to do something towards achieving this professional seamlessness in addressing the problems of the built environment. It aims to allow participants to cross boundaries, to explore and understand more fully their potential in contributing to the creation, maintenance and quality of our built environment.

With the Cambridge IDBE programme successfully launched, the Ove Arup Foundation was able to take other initiatives in furthering its objectives. Ove trained as a civil engineer. His interest in the formation of young people entering the profession centred largely on their understanding of design. His concept of design was holistic – a progression of concept followed by analysis and evaluation in an iterative manner. Most young graduates fresh from university see design more in the context of their general mathematical and analytical studies, which have always been, and continue to be, the foundation of engineering education.

There is a healthy debate about the different routes in the formation of the professional for the built environment. There are the evangelists who espouse the virtues of common academic experiences and programmes for undergraduates. Others feel equally passionately that the sheer quantity of material necessary for a proper education, let alone the appropriate rigour in most subjects in which students have to become proficient, makes it desirable to retain unidisciplinary studies. That is not to deny the value of contextual and interdisciplinary studies, particularly at undergraduate level or more commonly in postgraduate programmes.

The argument is complicated by the question as to how long an honours degree programme should take. This debate is particularly relevant to engineering degrees, but similar arguments abound in other disciplines working in the built environment. There are the pundits for four-year honours degrees. Honours degree programmes for engineers in this country appear to be shorter than those in competitor countries. There are legitimate complaints by academics that there are ever-increasing demands for more and more subjects to be introduced into undergraduate courses already crammed so full as to make it virtually impossible to give students a rounded education, let alone allow them to partake in some of the extra-mural activities which should form part of a university experience. Controversially this issue has now been resolved, at least for the time being, by the Engineering Council's direct route towards the chartered qualification via a four-year study programme. Three-year degree courses continue to produce engineers with appropriate standards for incorporation. There were pressures for two-year degree programmes, presumably in the context of improving productivity in our higher education institutions.

There is probably room for all these routes if we could only define our needs a little more clearly. After all, our industry is many-sided and needs different skills and talents at different levels. There is a danger of trying to homogenize the formation of professionals. What is really needed is a rich variety of talents, abilities and training. For instance there is some indication that the number of engineers in construction is not at issue. On the other hand it does appear that many engineers carry out tasks which are more suitable for technician engineers and technicians.

It might be helpful to look at education and training as a continuum not only in time, but also in content, so that a two-year degree graduate would not normally be expected to attempt tasks of the complexity expected from a person with a four-year degree or from one who has taken part in an appropriate postgraduate programme. But whether one is speaking about two-, three-, four-year or even postgraduate programmes, what is essential (and all too often missing), is a level of contextual understanding, of cross-cultural studies which are so necessary if our designs are to be properly integrated.

Design, in the widest interpretation of the word, is the confluence of all knowledge which engineers acquire. Ove Arup felt passionately about design as the underpinning and sustaining element of our built environment. He frequently wrote and spoke about its all-pervasive importance. In his 1970 Alfred Bossom lecture at the Royal Society of Arts he said *inter alia* that '...design is a mental activity devoted to figuring out and deciding how to make or build an entity, what it should be made of, what it should look like, how it should be made etc...'.

There are strong indicators that design, the compounding of flair and the students' acquired knowledge, is not being given sufficient emphasis in many faculties of engineering at our universities and that real design experience in civil engineering schools is lacking.

The need to expose students at least to some of the often conflicting circumstances and considerations involved in the realization of a project cannot be emphasized enough. Competing social, environmental, technical, economic and aesthetic demands must be understood if they are to be tamed and integrated into a design. Process and product are part of a seamless continuum. The culture of design as a creative process is, or should be, overarching.

There is *some* creative flair in all of us. It comes naturally to some; it needs to be teased out of others. Not all engineers will, or even should, become great designers any more than all doctors should aspire to become great surgeons, but the need for an *awareness* and *understanding* of the value of design is a general imperative. Whether as constructors or consultants, whether as academics or public servants, engineers, civil engineers in this case, aspiring to roles of leadership must understand design and have an appreciation for it. Even were they to leave the profession to work in the City or in politics or some other walk of life, an understanding of design can only be of benefit to them and the activity on which they are engaged and the public at large. It is only by active participation that concepts of process and product in the context of design can be understood.

Medical students acquire most of their clinical education from practising doctors and consultants, usually the best and most distinguished in their particular speciality. Unusually, engineering and particularly civil engineering students are rarely exposed to the leading practitioners during their degree programme. To give students an understanding of design should ideally involve a practising designer – competence in and understanding of design requires continuous involvement. There is an overwhelming case for civil engineering students to be exposed to and be guided by practising engineers, particularly those thought to be the most able.

These deliberations prompted the Trustees of the Ove Arup Foundation to sponsor a Chair of Civil Engineering Design at the Imperial College of Science, Technology and Medicine which is part of the University of London. It is a natural development of the admirable Visiting Professors Scheme launched and sponsored by The Royal Academy of Engineering. However, at Imperial College the incumbent takes responsibility for teaching design at undergraduate level. The proposition is based on a distinguished practising designer attending

College for sufficient time to plan, prepare and personally direct studies and the practice of design and interact with the students. The emphasis is on the post being led by a distinguished practising engineer, whose main activity is in designing all kinds of civil engineering projects.

All indications point to this initiative being highly successful. The enthusiasm which has been generated will, it is hoped, help to dispel the often held perception among school leavers that engineering education is not only difficult, but also dull. Undergraduates are given an added contextual dimension to their studies as well as an opportunity to express their creative flair. It remains for engineering departments in other universities to establish similar chairs. The sums of money required to sponsor such a chair are comparatively modest – the impact on the students and consequently on their contribution to their role, whatever it might be in the construction industry, could be substantial.

No aspect of the built environment currently holds greater challenges than the physical and social structure of cities. The movement of the world's population to cities has been continuous for two centuries, and unabated global urbanization is inexorable. To make cities function, to cater for the physical, economic and intellectual needs of their inhabitants, has set immense problems to all those involved in their planning, maintenance and government. The need to control and regenerate dereliction, to satisfy housing, education and other societal requirements of the burgeoning urban population pose unprecedented challenges. There are many theories about how to improve our cities and there are even some isolated successes in applying them. But these are all too patchy and the challenges posed by the issues of urban design in the full meaning of the word are arguably the toughest to be faced by all the specialists involved trying to work across disciplinary boundaries. To integrate social science, economics and politics with architecture and engineering into the contemporary cultural and environmental urban context is a huge challenge.

It was for that reason that the Ove Arup Foundation enthusiastically embraced the concept to help launch a postgraduate programme at the London School of Economics which would endeavour to bring together all those involved or participating in the design, maintenance and evolution of cities. The programme, which is studio-based, brings together graduate students whose first degrees range from architecture and engineering to economics and sociology. The aim of the programme is to establish an international community of urbanists able to influence seminally the improvement of urban life. This innovative approach to urban studies is now a flourishing centre alongside other long-established centres of excellence at the London School of Economics.

Some years ago a friend gave me a book written by members of his family. It is called *Making Connections – Teaching and the Human Brain*. It is an erudite work on teaching and learning which examines research in the neurosciences and its relevance to education. Much of it was beyond my understanding, but there was one quote which I did understand. A chapter which explores the mystery of meaning begins with this quote:

> *Two stone cutters were engaged in similar activity. Asked what they were doing, one answered, 'I'm squaring up this block of stone.' The other replied, 'I'm building a cathedral.' The first may have been under employed; the second was not. Clearly what counts is not so much what work a person does, but what he perceives he is doing it for.*

This is a very simplistic illustration of the problems we are facing in designing and maintaining the built environment. It is not only necessary for us to know what we are doing (and that is clearly quite difficult at times) but also why we are doing it. When we understand the 'why', the possibilities of new insights can, and often do, arise.

The initiatives taken by the Arup Foundation try to address the problem on a number of fronts. The Masters programme for interdisciplinary studies at Cambridge creates opportunities for all its participants to gain a fuller understanding of the 'why' in design – creative involvement in the 'how' in design usually follows. This should make an immensely important contribution to our industry. The Cambridge programme seeks to place emphasis on design and its underpinning knowledge. It is open to architects, engineers, constructors, surveyors, clients – in other words it is possible for all those participating in the process of creating the built environment to have common learning experiences. The Chair for Civil Engineering Design at Imperial College brings to undergraduates new outlets for their creative talents and the possibility of a broader understanding of the context in which they are working. The Cities Programme at the London School of Economics uniquely bridges the cultures of the sciences and the humanities in so far as they impinge on the functioning of urban life – interdisciplinary working at the highest and most challenging level.

These and similar educational initiatives are necessary to help solve today's problems and to plan for coping with tomorrow's. It is no longer enough to understand our own individual specialization. We need to have an understanding of the whole not merely the part.

H. G. Wells suggested rightly that 'Human history becomes more and more a race between education and catastrophe'. Much of our history is now inextricably linked with and influenced by our built environment. Interdisciplinary studies for the built environment are at the very core of aspirations for sustainability and survival let alone improved living standards.

**CHAPTER NINE**

# Experiences of interdisciplinarity: observations from the Masters course 'Interdisciplinary Design for the Built Environment', Cambridge University

*Paul Kirby*

## Taking up the challenge

If 'a camel is horse designed by a committee', we had better decide whether the built environment should come with one lump or two. Or we should build a better committee. Any project of a scale that has an impact in the public realm will involve the knowledge, skills and judgement of more than a single creator and in consequence design is a collaborative exercise. Our committees, or design teams as we more normally call them, had better learn to design well. This calls for designers who can define a good result and can work together to deliver it.

Both the Latham and Egan reports to government (Latham, 1994; Egan, 1998) highlighted fragmentation as a cause of conflict and inefficiency in the construction industry, but before either of these were published the course in Interdisciplinary Design for the Built Environment (IDBE) was inaugurated in Cambridge with the ambition of building a better understanding between the professions engaged in construction. This chapter illustrates the philosophy of the course by drawing on some of the experiences gained in running it and discusses also the nature of 'interdisciplinarity' itself. But first a potted history.

In September 1991, as mentioned by Sir Jack Zunz in the previous chapter, The Ove Arup Foundation sponsored a conference[1] in Cambridge which sought to suggest ways in which education might

1 The conference proceedings were privately published as *Education for the Built Environment*.

encourage a greater appreciation of the multi-disciplinary nature of design. One outcome was the award by the Foundation of startup funding for an interdisciplinary Masters degree. At the same time the University of Cambridge was drawing close to establishing part-time Masters degrees which it had never previously offered. Professor Peter Carolin, at that time head of the Department of Architecture, saw the opportunity that these two initiatives offered and proposed, in competition with other institutions, the framework of the course that subsequently became IDBE.

Before going further I should perhaps point out that the choice of a Masters degree rather than an undergraduate degree or undergraduate course module does not imply that this is the only model that interdisciplinary education should follow. Undergraduate degrees that include subject matter beyond the traditional boundaries of their discipline certainly encourage an appreciation of the collaborative nature of design. There remains, however, a strong case for a postgraduate course, as young professionals, having spent a few years in practice, will be better placed to embark on broader areas of study than raw graduates whose first need is to establish competence in their home discipline. The more mature student will have first-hand experience of practice, its opportunities and complexities and will be able to discuss, with knowledge, issues which at undergraduate level can be raised only as abstractions.

The IDBE course is therefore seen as an adjunct to, rather than as alternative to, interdisciplinary exposure at undergraduate level, and it would be disappointing indeed if Masters level education were required simply to rectify defects brought about by the tunnel vision of exclusive specialism at this earlier stage.

The IDBE course enrolled its first students in July 1993 and by September 2000 had attracted a total of nearly 100 students. As intended, these students have been drawn from all of the main built environment disciplines. Figure 9.1 shows how various disciplines are represented

The part-time structure of the course brings the students together for week-long study periods in Cambridge. During these weeks the students attend lectures and seminars, but also take part in short, sharp design exercises. What distinguishes these exercises from most design education at undergraduate level is that it is undertaken in multi-disciplinary groups. The design projects thus become a vehicle for exploring both the design of the product, for example a generic laboratory building or a small urban settlement, and the design process itself.

The course is essentially about the practice of design and seeks to help create effective, practising designers. Since this is the aim, a group

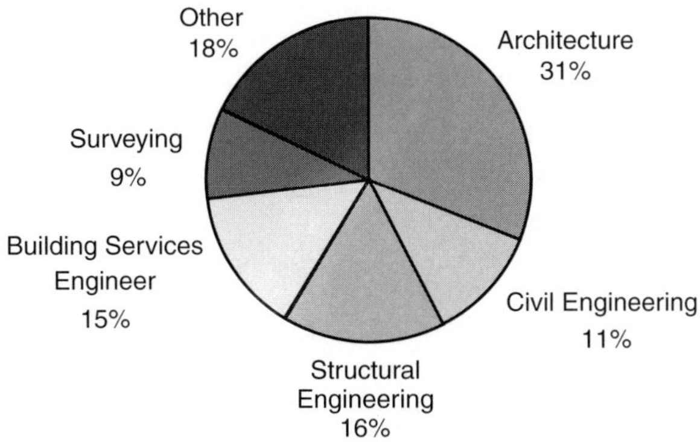


*Figure 9.1 The professional background of IDBE students. Within the category 'other' are students from, for example, landscape design, product design and manufacturing, and from disciplines with a more remote professional connection to construction.*

of employers[2] were asked what, in their opinion, were the qualities needing to be developed in their designers. A summary of their comments is shown in Figure 9.2.

It is notable that nowhere on this list do 'greater technical competence' or 'refined specialist skills' appear. The greatest gap seems to be in identifying solutions and mobilizing collaboration. Notably, all of the qualities listed could apply to all disciplines and closely parallel the objectives of the IDBE course. All could be thought to represent qualities that the interdisciplinary designer should aspire to.

## Cultures and differences

Employers paint a picture of a need for inventive individuals who are capable of mutual understanding, Latham and Egan are concerned by fragmentation, and more generally the industry is commonly represented as riven with conflict.[3] We might say that lines of communication are breaking down due to rapid advances into specialist purism, but specialization is beneficial. How then are we to describe difference in a

2 Though it was a small sample (six individuals from five organizations), the consistency of the replies gives some confidence that the views are widely held in the industry. Interviewees were principals, senior executives or directors of their practices, and these included RMJM, BDP and Ove Arup and Partners.

3 Are there any other trade journals besides *Building* which have contained a regular feature called 'Who is Suing Who'?

| | |
|---|---|
| **Leadership**<br>Learning to act through others<br>Appreciating and judging the contributions of others<br>Controlling the design process<br>Understanding the roles of all participants<br>Managing teams | **Client relationships**<br>Understanding client objectives<br>Understanding financial structures<br>Project management<br>Briefing |
| **Problem solving**<br>Creating alternatives<br>Team working<br>Clarifying objectives<br>Balancing price and quality<br>Designing with confidence (esp. engineers)<br>Awareness of cultural change | **Personal skills**<br>Breadth of vision<br>Communication skills<br>Personal adaptability to change<br>Holistic vision |

*Figure 9.2 A summary of views expressed by a sample of employers when asked by the author what skills need to be developed in young designers.*

way that helps us to identify which differences are destructive and which are constructive?

A first approach might invoke a reconciliation between C. P. Snow's 'Two Cultures' (Snow, 1959). The simplicity of a humanities–sciences division is quite seductive, and though such a comparison may offer insights (and still has some currency[4]) Snow could not perhaps have envisaged the further proliferation of specialisms, sub-specialisms and special interest groups which populate contemporary society. The humanities–science divide cannot even capture the full diversity of specialisms and value systems that are abroad within construction. Here we find the entrepreneur, the designer, the constructor, the manufacturer, the user and the passer-by, and no simple partition can separate these groups exclusively into one typological camp or another. All have particular skills and inclinations that differently configure their understandings and capacities.

Within the small and self-selecting group of IDBE students a further measure of diversity is to be found in the differing motivations of the students. These differences offer shades of meaning to interdisciplinarity itself. I do not pretend that the six categories below are authoritative, but they arise from observation of the students and I would be surprised if they could not be recognized also in most offices.

4 The organization of the reading rooms in the new British Library perpetuates this division.

First, we have the competitors, whose faith in their own ability causes them to be frustrated by the apparent lack of competence within other disciplines and who wish therefore to wrest some power from them. Second are the comfortable specialists who seek to enrich their home discipline by exposure to new stimuli. Third are team workers, wishing to understand their co-collaborators more fully and wishing to be heard more clearly. Fourth are team leaders, needing to understand the motives and methods of others in order to be more able to direct them. Fifth are the renaissance princes and princesses with ambition neither to lead nor be led, but rather to fulfil a quest for personal growth. Finally, and in smaller numbers, we see on the course the disenchanted. This group has lived out the consequences of a premature choice of career and find that the consequent enforced specialization leaves them professionally competent but uninvolved.

None of the students apart from the few who I term 'the disenchanted' wish to break away from their core discipline; rather they recognize that becoming more competent in their discipline requires that they must expand their vision beyond the historic professional boundaries. Neither does the course attempt to undermine specialization, as it assumes that the knowledge that specialization brings remains essential if the complexity of design projects is to be safely navigated. It is also based on the understanding that originality of viewpoint is essential to innovation and indeed ordinary levels of inventive problem solving.

Difference is inevitable, it provides necessary tools for analysis and it is a stimulus to inventiveness: but it can also have a disintegrating effect when stereotypical behaviours inhibit collaboration. Stereotypes must always be generalizations and are usually unhelpful, but there is an observable tendency for all disciplines to retreat to familiar positions in which they feel comfortably competent. We see quantity surveyors measuring, engineers calculating and architects drawing details when their collective minds were better engaged in imagining a solution to a shared problem. When the going gets tough the challenge of collaborative imagination may be more useful than the comfort of specialist routines.

Similarly, the conventional division of thinking into convergent and divergent types can be unhelpful when it lends plausibility to such stereotypes. It is often said that architecture represents divergent thinking, that is thinking which seeks out alternatives and explores the 'solution space', whereas the numerate disciplines are characterized by convergent thinking, that is they tend, by a process of logic or algorithm, to work towards a one and only possible solution. While these modes of thought may exist, it is extremely unhelpful to then assume that one mode of thinking should be the sole prerogative of one or other discipline. Design is an iterative process in which convergent and divergent thinking play their part in loops of invention and analysis[5] and it is destructive to allocate modes of thought according to disciplines rather than the imperative and dynamic of the design task. The danger is not that the disciplines falsely ascribe unique qualities of thinking to their peers, but that they accept the faulty description of themselves. By undervaluing, and thereby neglecting to develop, the full range of their talents, designers may diminish their ability to contribute to their full potential.

The symptoms of this (as observed on the course) are that some students distrust a systematic analysis of physical constraints because of a perception that logic and mathematics grinds relentlessly towards an inevitable solution, and that this inevitability threatens the role of individual creativity. Conversely, others rely too heavily on analysis while failing to develop their ability to invent a model to analyze. There seems to be a continuum, with self-expression at one extreme and the expression of abstract universals at the other.

An extreme example from the course illustrates this. An early project given to IDBE students is the design of a facade system. The project encapsulates technical, ergonomic and aesthetic criteria in a formidably complex interdisciplinary design task. One design team spent the time allocated to the project endlessly calculating alternative strategies for natural ventilation, solar gain, daylighting and so on, with the intention of formulating the perfect solution (the abstract universals of building physics were believed to hold a solution within them). They were chastened to find that no solution emerged. Internal conflicts within the technical criteria prevented it, making choice and judgement necessary. It was further pointed out to them that not only had they arrived at no 'technical' solution, but that, by not beginning to grapple with the design of a physical object they had not even begun to address the criteria that relate to the user and how these might introduce further conflicting criteria. This 'convergent' group needed to learn that the initiation of a design solution is not the outcome of analysis but of imagination. As Karl Popper says, 'there is no such thing as a logical method

5 The IDBE-linked research project 'Mapping the Design Process' soon revealed the iterative non-linear nature of conceptual design in which speculation is the tool of advancement. That is, a model (idea) is constructed to see what works and to investigate unknowns. It is rarely possible to encompass all necessary criteria and opportunities at the outset and to condense a solution which is fully formed in its first incarnation.

of having new ideas, or a logical reconstruction of this process' (Popper, 1934).

Styles of thinking, cultural backgrounds, education, motivation and any number of other influences create, in turn, a diversity of purposes, methods and perceptions. In discussing the role of judgement in engineering, Parkin considers how such differences lead individuals to form different judgements about the same data and illustrates this by means of a metaphorical lens (Parkin, 2000). Figure 9.3 shows how the perceptions of two 'judges' viewing the same problem or opportunity will vary according to each judge's particular sensitivities. These in turn are derived from education, values and the dominant influences of the 'subculture' that each designer inhabits. The figure is a useful description of the familiar experience of conflicting understandings and more importantly of the need for there to be a strategy for focusing a design team's effort on a single point, where that point is a jointly conceived design solution.

Whether or not such a strategy is consciously conceived, a strategy will be implemented. The need for a result demands it. Some of these familiar default strategies have been played out on the IDBE course.

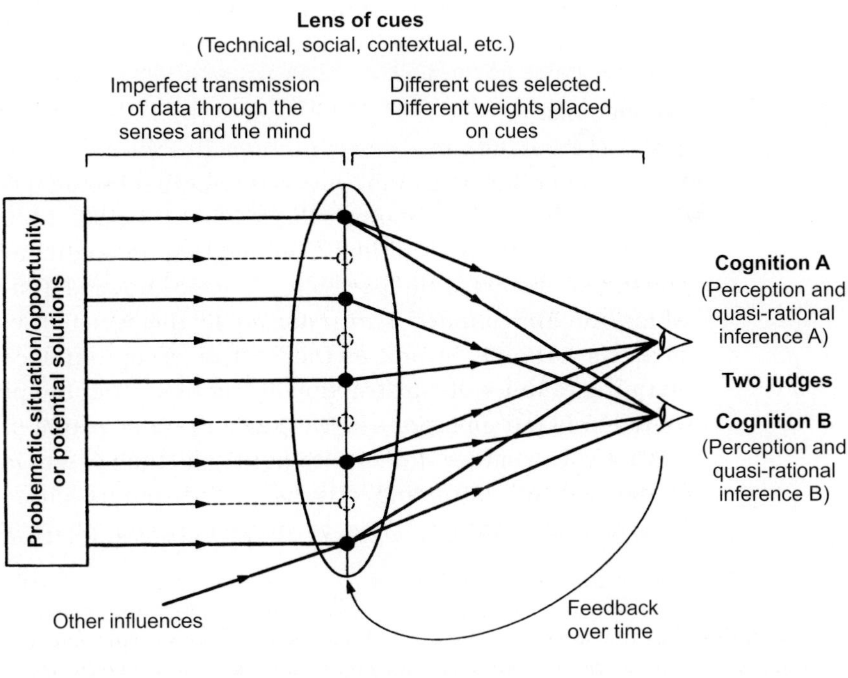


*Figure 9.3 A lens model of judgement (Parkin, 2000).*

## Flawed strategies

### Kit of parts

Take the facade design again as an example. Most teams segment the design task according to different groupings of criteria and bring their specialisms to bear on understanding the nature of problem. But some groups falter at this point as they cannot then imagine a mutually satisfactory, coordinated solution. Instead, the team remains broken into specialist groups which independently solve subsets of the problem. The reduction of the problem to soluble sub-problems could be said to be normal practice, but the coordination of the sub-problems into a workable whole is the difficult next step which these groups cannot make. So instead of jointly solving the problem they offer a conglomeration of partial solutions, the 'kit of parts'.[6] Thus, for example, glazing might be sized for daylighting and one of the 'parts' offered is a louvre, which the user can deploy to counteract solar gain. Unfortunately the conflict between light and heat is not solved, but is passed on to the user disguised as choice. Users are offered a spurious control over their environment in that they are able to choose between being hot and well lit or being cool in darkness. This is not a sensible choice. There is a role of course for 'kit of parts' solutions to design problems, but only if the choice of parts or the use of the parts actually solves the problem and is not a strategy for avoiding conflict by passing it on.

### Compromise

At the heart of the camel/horse problem lies the sinister form of compromise. When asked in a radio interview whether he could be tempted into politics, Mr Gerry Robbins, the CEO of Grand Metropolitan and Chairman of the Arts Council,[7] replied that he might have difficulty in practising 'the art of the possible' – a well-known definition of politics which for him implied surrendering to the worst of compromise. It suggests bargains struck at the cost of accepting second best. It is tempting to think of construction as another art of the possible, but do we too limit our ambitions to the second rate in search of compromise?

What construction shares with politics is the need to make decisions in the face of conflicting criteria and disparate interest groups. In construction, conflict tends to arise between physical criteria (for

---

6 The student design teams frequently use phrases such as 'kit of parts' and convince themselves that having named the strategy it becomes somehow legitimate.

7 Interview on BBC Radio 4.

example natural ventilation and noise) and between values (for example engineering integrity, commercial necessity and perhaps environmental concerns). In addition to the technical problem of resolving the conflicts is the problem that the different criteria may be defended by different specialists within (or outside) the design team. In this environment compromise may prevent a conflict but, if we aspire only to the least worst fit, does it produce good design?

### Horse trading

It would be true to say that agreements are easier to make if the starting points of the two (or more) parties are not too distant at the outset; however, whether the circumstances are arguing a final account or for space for structure, 'horse trading' is a common tactic. That is, each side overstates its needs in order to have spare capacity that can be traded away. The overstatement assumes conflict and, more destructively still, by polarizing the interest groups, creates it. Damage limitation is the intent rather than mutual benefit.

We can see these and other flawed strategies (for example bullying, pulling rank) employed to force a result, but because the strategy is flawed, flaws will remain in the design itself. Differences are not accommodated but are exaggerated, steamrollered or ignored.

Conflicting physical parameters or project objectives are an inescapable part of design, but differences in attitudes and values will also exist within design teams. These compound the difficulty of creating a joint decision by adding a layer of interpersonal misunderstanding and conflict. It should be noted, for example, that the 'kit of parts' strategy is an effective means of avoiding discussion and is frequently adopted by design groups in which there is a clear lack of empathy (verging on hostility) between the members.

## Value in difference

If we accept that differences can be destructive, an important choice becomes apparent. Do we choose to homogenize the design team so that differences are erased, or do we provide tools by which differences can be accommodated? If we accept the multiplicity of world views that cultures and subcultures engender, particularly in an increasingly international milieu, then the possibility of creating a single value system may be impossible to achieve, even were it proved to be desirable.

The IDBE course assumes that difference is both inevitable and valuable, and, while it may seem paradoxical, it is assumed also that it is a lack of insight into difference that is a barrier to achieving mutual understanding. That is to say, difficulties in forging agreement can arise not from a clear view of differing goals and methods but from the assumption that people are mostly alike and that conflict is due to an arrogant inflexibility on the part of others. To build shared understandings, therefore, it is first necessary to highlight difference and to demonstrate that by respecting difference conflict becomes understandable and can be negotiated.

One exercise that is introduced on the IDBE course is personality profiling.[8] By completing a questionnaire the students learn which of 16 personality types they most correspond to. Since each of the 16 suggests differences in attitudes and behaviours, difficulties in the relationships between students can be discussed in terms of legitimate differences in world view rather than inexplicable perversity and stubbornness.

Similarly, the students are introduced to the work of Meredith Belbin, who has published widely on the subject of the roles adopted by individuals within management structures (Belbin, 1993). He has identified eight roles which include, for example, 'Resource Investigator', 'Completer/Finisher' and 'Monitor/Evaluator', and suggests that individual character traits may lead individuals to take up particular roles in teamworking environments. The limitation of the Belbin approach is that it is most easily understood in terms of management groups within which individuals have similar basic skills and where differentiation of role is the product only of personality, whereas within design teams individuals are specialists who are forced by their specialism rather than their nature to take up particular roles.[9] Nevertheless, the students tend to find their own behavioural characteristics plausibly illustrated by Belbin and perhaps more importantly recognize that for teams to be effective the roles described need to be filled, whether or not the team is selected from a group who fill them naturally. What is absent by nature must be provided by deliberate action. No matter what the personal characteristics of a team may be, the team had better ensure that ideas are generated, information is sourced, progress monitored, and judgements are critically evaluated.

8 Using the categories developed by Briggs and Myers (1995). The personality types are not represented as an unarguable truth on the IDBE course, but are used to illustrate potential character differences between individuals.

9 Belbin's later work introduces a 'specialist' team role to act alongside the others. A design team, however, is comprised entirely of specialists. These have to fill the conventional team roles while simultaneously acting as specialists.

It is made clear to the students that Belbin's categories of teamwork roles are not offered as absolute truths and are not the only profiling system to be found among the techniques of psychometric testing. Nevertheless, and as for the Myers Briggs profiling, mediating the discussion of differences through typologies of this type helps to dissolve prejudicial stereotyping by legitimizing alternative attitudes and behaviours.

Discussing interdisciplinarity in terms of personality, thinking styles and team roles diminishes the role of knowledge in creating a collaborative environment and stresses instead the need for fostering appropriate attitudes – attitudes which include generosity, openness and inquisitiveness and which engender mutual respect. To return to Parkin's lens, we have to ask how these attitudes help define the interdisciplinary goal and how they might guide our disparate perceptions to this point of focus.

## Resolving difference

It is a pity that the word holism seems to have been co-opted by the troublesomely mystical practitioners of 'New Age' therapies. Pushed too far, ideas of holism challenge the value of 'reductionist' thinking; indeed 'reductionist' has become a mild term of abuse. This is a pity, as we depend upon the reductionism that specialization brings to provide detail to our analysis. In its broader, more permissive sense, however, holism defines the core ambition of interdisciplinarity, which is that all stakeholder requirements, technical criteria, constraints and opportunities should be identified, accommodated and harmoniously satisfied.

I should add at this point that the harmony should be in the product. The process itself might not be without conflict between individuals, but with interdisciplinary working comes the acknowledgement that such conflict can highlight conflicts within the design criteria and goals themselves. Clashes between disciplines can, if resolved through effective team work, prevent conflict in the product itself, as Bucciarelli (1994) says:

> *Each individual's perspectives and interests are rooted in his or her special expertise and responsibilities. Designing is a process of bringing coherence to these perspectives and interests, and fixing them in an artefact. Participants work to bring their efforts into harmony through negotiation.*

It can be readily observed, in both practice and as played out on the IDBE course, that design is the product of negotiated agreements over what is to be done. Design is not simply the coordination of solutions to discrete sub-problems, but a synthesis brought about by the negotiation of mutually satisfactory outcomes. Aiming higher than compromise, such a synthesis does not set one group of criteria in opposition to another, but seeks a gestalt or synergy. This reality tends not to be made explicit by design flow charts,[10] which deal with actions, nor by information flow analysis, which tracks documents rather than decisions. Agreements negotiated between clients and design teams and between specialists within design teams are the stuff of design. Interdisciplinary design accepts this and seeks to make negotiation fruitful.

In one workshop on the course negotiation is discussed explicitly. The workshop is based upon research undertaken at Harvard University which is helpfully summarized in the book *Getting to Yes* (Fisher and Ury, 1997). One of the messages of the workshop might be summarized as: success in negotiation comes not from forcing your opponent to give ground but from making it easy for them to agree with you. How might we do this in design?

On the IDBE course the principal answer is to understand the preoccupations and methods of the other members of the design team. This understanding enables the designer to frame his or her part of the solution in terms which already address the needs of other specialisms. In accepting the relevance (and authority) of other criteria, the designer avoids creating partial solutions that in their single-mindedness engender unnecessary conflict. The return for taking the time to broaden one's own understanding is that, in a suitably interdisciplinary environment, those with whom you are forging agreements will themselves be attempting to meet your design needs.

As an aside it could be argued that a step beyond a sensitivity to the criteria of others is a preparedness to contribute directly to elements of the design considered to be within another discipline's domain. For example: the client may suggest structural solutions, the architect may comment on the client's organizational structure, and the building services engineer might offer suggestions for disabled access. Good ideas can come from any source.[11] But if design is not to degenerate into an ill-informed free for all, then those who seek to widen their design horizons must be sufficiently knowledgeable of the nuts and bolts of other disciplines to be able to make sound suggestions and must have

10 For example the widely quoted example from Pahl and Bytes.

11 Lewis Wolpert, himself a civil engineer who became a biologist, offers some interesting reflections on the subject of creativity in *The Unnatural Nature of Science* (Wolpert, 1992). See also Sam Price's comments in Chapter 6.

the good judgement to know the difference between contributing and interfering. Assuming the necessary level of raw intelligence, such knowledge and judgement will come with experience. With younger designers taking greater responsibility earlier we have to ask, however, whether there is enough time for this experience to be accumulated. The IDBE course offers an alternative, in that the students' studies can bring a rapid and deep exposure to other disciplines and allow the type of experimentation which the pressures of time and litigation impede in normal practice.

Productive negotiation requires a generosity of attitude. It needs an honest willingness to accept the validity of criteria other than one's own. Where there is conflict, however, there exists the countervailing pressure for team members to act as advocates on behalf of, say, structure, or cost, or space. This gives scope for the more able advocate to promote the cause of his or her home discipline over the legitimate but less confidently argued interests of other disciplines. In place of collaboration we see a battle of wills, where the acceptance of another's point of view becomes a weakness to be exploited. When the defence of valid criteria becomes an attack on and defeat of an opposing engineer, architect or cost consultant, for example, the project may be deeply wounded. Negotiation and interdisciplinarity need to break free from ideas of advocacy, and individuals need to conceive of themselves (and others) not as the champions of their discipline but as contributors to the whole, having particular and essential expertise that the group can draw upon.

Good timing is also a necessary condition for good agreements. This is why the early involvement of all relevant disciplines is essential to interdisciplinary working. The alternative and almost universal practice is for one or other of a design team to be appointed ahead of the remainder. The unfortunate outcome is a tendency for a polarization in the design team, as unless the first person on the scene is prodigiously multi-talented, preliminary thinking will embody biases in the design criteria which later contributions will need to re-address. Opportunities will be missed, and the need for compromises established long before the problem is even fully digested by the team. Argument (no matter how well mannered) will replace negotiation as the design team struggles to oppose or defend uncoordinated design initiatives.

The importance of these first moves, the strategic decisions, is illustrated by the familiar chart showing opportunities for change declining through the lifetime of a project as the cost of changes increases. The implication of this, if it is accepted, is that a small improvement in the quality of early decisions will have a far greater effect on project success than improvements brought about during design development. If we additionally accept, as argued above, that conceptual design requires coordinated thinking, then we must accept the potential for early interdisciplinary thinking to offer substantial reward within the overall design process. For this reason IDBE design projects concentrate on these early strategic decisions.[12]

While the knowledge of other disciplines that can be transmitted by a Masters course must be limited, this may not necessarily be a cause for concern, as few designers bring the full weight of their specialist analytic tools or repertoire of techniques to early design proposals. What they bring, apart from an inventive mind, are experiences of what they know to work and a few rules of thumb. What a Masters course can do is to expose students to some of this knowledge (through design exercises and case studies) so that while they will never be accomplished outside of their own discipline, they will be able to share a common understanding of multiple design criteria, will be able recognize opportunities which can bring multiple benefits and can create starting points from which a shared solution can be negotiated. They may not have the capacity to analyze these in detail, but that, after all, is the role of the specialist.

## To what end?

At the beginning of this chapter interdisciplinarity was offered as a means of defining and achieving good results. Real difficulties arise, however, in the definition of the good result. Poor results are well illustrated in the Latham document, which includes a comparison of clients' views of manufactured products and buildings.[13] Buildings were seen to perform relatively poorly, suffering from late delivery, defects and cost overruns. The value of Latham and Egan was the presentation of these symptoms not as a regrettable but perhaps inevitable product of a fragmented supply infrastructure, but as an unacceptable denial of the rights of the demand side. This re-focusing of attention on the client revitalizes discussions about quality by grounding the discussion in the

12 We repeatedly hear engineers complaining that they are not invited to contribute to the design early enough, but tellingly, when, on the course, engineers are invited to contribute, they frequently adopt the conventional position of 'waiting for something to analyze'. In consequence, architects can with some justification claim that when invited to contribute engineers are often unable or are not disposed to do so.

13 The Latham report quotes material presented by Dr Bernard Rimmer of Slough Estates plc which suggests that industrial and commercial buildings in particular have a poor reputation for the quality of construction, durability, fitness for purpose and protection by guarantees.

realities of the customer's experience. It is not sufficient for the design team to determine for themselves and in their own terms what constitutes good or appropriate building. This is not to suggest that design teams have traditionally acted in bad faith towards their clients, but it must be accepted that the supply side had shown little inclination to promote targets for cost reduction and quality improvement of the type demanded by Latham and Egan. Had it been so, current industry initiatives – M4I, construction best practice programme and so on – would have been superfluous.

We might ask, however, if the client is king does that make the rest of us serfs? If designers, as they are exhorted to do, place themselves in the position of being the agents through which the client's will is done, what is to become of the professional's historically assumed duty to the wider public good? It might be dangerous to assume that benefits to clients aggregate into benefits for all. Do decisions based on the 'business case' guarantee a decent environment for the user? On the larger scale we might ask whether the property market could be said to be sensitive to the ambitions of sustainable development. This is not the place to discuss this topic at length, but when grappling with definitions of quality the debate must consider, 'Who is the arbiter?'.

The IDBE course does not seek to answer this directly, but does expose the students to a variety of sources who have an interest in the solution. Clients are represented both in person and through an examination of client-led initiatives. Users are brought into the picture by introducing, for example, feedback from the Probe Studies,[14] as are broader communities by introducing the students to lectures by practitioner/commentators such as Harley Sherlock, 'Cities are Good for us', Nick Wates, 'Community Planning', and Alan Baxter 'the Civic Engineer'. Beyond such direct exposure students also explore through project work (under the title 'Infrastructure and Landscape' for example) broader issues such as the conflict between the different value systems represented by our need for both construction and environmental protection – a dilemma that is not solely represented by the clashing of differing interest groups but which may be present as a conflict within each of us as individuals. Even those dedicated to construction will have a value-laden back yard that they wish to protect.

The concept of sustainable development has many shortcomings, not least a lack of clear targets. A first question might be 'What is it that we wish to sustain?', which itself leads to questions concerning quality of

14 Adrian Leaman and William Bordass for *Building Services Journal*.

life. What do we value and what do we aspire to? For some, sustainability means a reversion to subsistence; for others it means little more than business as usual, with a few token but high-profile nods to lower energy use. The value of the idea, however, is that it defines success in terms of maximum utility from minimum resources, and this sits very comfortably within the definition of good design.

Sustainability is also discussed in terms of three sets of criteria: economic, environmental and social, which development needs to satisfy. Again, it is not difficult to see how designers might readily adopt these criteria. The built environment is naturally a product of and a contributor towards 'economic, environmental and social' well-being.

Sustainability is therefore both a desirable end in itself and also a useful tool for examining the purposes and values which lie behind design. It also enlarges the constituency whose voice must be heard, beyond those who commission and design buildings, to include those who are not formally part of the decision making processes but yet who enjoy or suffer from the consequences. Consultation is a strong theme within Local Agenda 21, for example, and designers may well face a future which demands that they not merely learn to accommodate each other but listen to and negotiate with a more fully participating public.

The difficulty is that the breadth of the challenge is compounded by the breadth of the technical problems posed and, importantly, their interdependency. For example, is a low-energy green field office building (which depends upon private transport) more or less sustainable than an air-conditioned office on an urban brown field site? But it is these interdependencies within sustainability that both illuminate the need for interdisciplinarity and provide a tool with which to examine it.

## Conclusion

Are there an engineering mentality, an entrepreneurial mentality, and an architectural mentality? If there are, we could ask whether they are innate, formally taught or are acquired through identification with a peer group. Perhaps research could tell us. What is clear, however, is that behaviours and attitudes are observable which are consistent with there being such differences, and what is also clear is that these differences can be valuable. For our design committee to succeed we need to build a respect for these differences, we need to negotiate towards new opportunities and away from compromise, and we need a shared ambition. This, let us hope, is a built environment which future generations may inherit with gratitude.

## References

Belbin M. (1993) *Team Roles at Work*. Butterworth Heinemann, Oxford.

Briggs, I. and Myers, P. (1995) *Gifts Differing*. Davies-Black, Palo Alto, CA.

Bucciarelli L. (1994) *Designing Engineers*. MIT Press, Cambridge, MA.

DETR (1998) *Rethinking Construction: the Report of the Construction Task Force*, July 1998.

Fisher, R. and Ury, W. (1997) *Getting to Yes*. Arrow Business Books, London.

Latham, M. (1994) *Constructing the Team. Final Report of the Government/Industry Review of Procurement and Contractual Arrangements in the UK Construction Industry*. HMSO, July 1994.

Lawson, B. (1980) *How Designers Think*. Butterworth Architecture, Oxford.

Parkin, J. (2000) *Engineering Risk and Judgement*. Thomas Telford, London.

Popper, K. (1934) *The Logic of Scientific Discovery*. Hutchinson, London.

Snow, C. P. (1959) Rede Lecture, reprinted in *The Two Cultures*. Cambridge University Press, Cambridge, 1993 (with an excellent and extensive introduction by Stefan Collini).

Wolpert, L. (1992) *The Unnatural Nature of Science*. Faber & Faber, London.

## CHAPTER TEN

# Working together: short courses for design teams

*John Steele, Andy Parker and Paul Kirby*

### Design in practice: the individual versus the team

Designers, within construction and other industries, are finding a need for collaboration and are forming teams which span departments, companies and disciplines (Frankenberger and Auer, 1997). This coming together is driven by the scale and complexity of problems that designers are currently expected to address. Teamwork can assist in taming these problems by enabling diverse expertise and the division of labour to be applied. However, the benefits afforded by the presence of a multitude of viewpoints can quickly evaporate if the individual components of the design team do not cooperate. In fact, the difficulties faced in addressing problems can be compounded by the introduction of incompatible individuals, goals and methods. In these conditions the introduction of teams may only add to the wickedness (Rittel and Webber, 1973) of the problem. It is clear therefore that the disintegrating effect of difference needs to be counter-balanced by giving attention to the deliberate and intelligent integration of effort. This in turn requires that designers are equipped with a set of skills and attitudes that, regrettably, their tightly focused professional education may not have developed.

This chapter describes how one organization, AMEC, is seeking to correct this deficiency. It introduces the purposes and methods of AMEC's 'Designing Together' workshop and provides some assessment of its role and value to the organization and its members.

### AMEC: the organization

AMEC provides services and engineering solutions to infrastructure, manufacturing and process industries worldwide. The group believes that it sets itself high standards, its philosophy being that contributing to the long-term success of clients contributes also to AMEC's own success. In pursuit of this ambition it seeks to develop and deploy not only the necessary technologies but also appropriate management and organizational environments.

The organization as a whole comprises three parts: Services, Investment and Capital Projects, involving the company in the financing, design, construction and operation of infrastructure and buildings. It is a spread which provides AMEC with a comprehensive understanding of the short- and long-term needs of clients and which has led to the development of its multi-disciplinary capabilities. The multi-disciplinary approach is particularly strong within AMEC Capital Projects, the division with responsibility for the design and project management of buildings for the infrastructure and services sectors. In these fields clients involved in the transport, pharmaceutical and chemical fields are particularly well represented.

Since its foundation, originally as IDC, the company has recognized the advantage of maintaining a knowledge and skills base in all the principal design and management disciplines. However, the company also recognizes that the creation of a multi-disciplinary workforce does not, in itself, guarantee multi-disciplinary working practices. What are required, beyond simple co-location and a common management structure, are systems and attitudes which promote a truly collaborative environment. Extending this ambition beyond its own walls, AMEC is also intent on establishing more productive engagements with outside parties, including clients and suppliers, and is moving away from traditional (and traditionally adversarial) contractual relationships towards models that embrace partnerships, alliances, and framework agreements.

The technical and organizational complexity of the projects and this ambition towards more effective collaboration require staff who are not simply competent in their own discipline, but who are able also to communicate productively with their colleagues within and beyond AMEC. It is a skill rarely available 'ready to use' from school, college or university, and it is for this reason that AMEC has a history of investing in the education of its staff, the subject of this chapter.

## Training

The company is energetic in the development of its employees and looks beyond routine attendance at professional institution events. Each individual has very different training needs, and AMEC has found that these are best served by allowing designers to identify for themselves their strengths and weaknesses and create a corresponding development agenda. The progress of employees is monitored through regular reviews, and at these meetings employees from all disciplines have consistently reported their desire for a better understanding of the needs and objectives of their fellow teamworkers. The majority of the design disciplines have never been exposed to such interdisciplinary perspectives, and the assumption that designers understand one another, seems, therefore, ill-founded (Parker and Steele, 1998).

It is AMEC's philosophy that such a common understanding of values is essential to multi-disciplinary teamworking, as a lack of empathy with the goals of other disciplines promotes the raising of barriers and an inhibiting confrontational atmosphere. It has been shown that the negative effects on performance resulting from adversarial relationships within a workgroup can be both extensive and far-reaching (Austin *et al.*, 1999).

AMEC recognized a natural alignment between the direction of training requested by its staff and its own perception of the abilities seen as fundamental to its effectiveness. One of the responses to this was the development of a series of focused training exercises, the Designing Together workshops.

## Origins of the workshops

The format of the Designing Together initiative was derived from a similar workshop forming part of the Interdisciplinary Design for the Built Environment (IDBE) Masters course at the University of Cambridge. The second named author, who was a member of the first IDBE cohort (in 1996), recognized the potential of the workshop environment as a means of allowing designers to better understand their colleagues while working as part of interdisciplinary teams. On completing the IDBE course, he proposed to the AMEC board that a training event built around the IDBE workshop might run internally. This proposal represented the platform on which the AMEC Designing Together Workshop (DTW) was built.

The IDBE workshop was a single component in a Masters degree programme and was therefore integrated into, and synchronized with, a number of related events and teaching sessions. In this form it was a complementary constituent of a larger curriculum. Given that the AMEC version was required to stand alone, and was not linked to a teaching programme, it was not intended that it would teach specific technical content, but would instead enable the participants to investigate their roles within the design team. Changes were therefore made to tailor the workshop to the specific needs of the workplace environment.

## Objectives of workshops

As the academic theory of interdisciplinary collaborative design is limited, and the majority of what is commonly known about design comes from studies of individual designers (Cross and Clayburn-Cross, 1996), the Designing Together Workshop does not attempt to transfer a particular and well-established body of knowledge. Neither does it aim to prescribe any form of standard team working methodology. Instead, it offers the delegates a forum within which to interact, learn from one another and explore the processes of design while free from the rigours and pressures of the office environment. It also enables those involved to learn about themselves, while providing an opportunity to practice management and presentation skills learned on other courses. In short, the objectives of the course are to:

- initiate an interdisciplinary design culture
- provide an environment where designers can practice innovation without risk
- increase the standard of design integration
- promote the concept of 'seamless' design
- enable participants to understand how to work with a range of disciplines.

In offering the delegates an opportunity to examine these issues for themselves, the workshop encourages them to participate in the creation of strategies for collaboration. In place of offering rules and methods it builds an awareness of the need for designers to direct their intelligence towards the process of designing (as part of a team) over the long term. The workshop offers insights and questions that the designers may further elaborate (as reflective practitioners) while actively designing in the real world.

## Attendees

The workshops, which are held over a two-day period, each involve 15 delegates. These delegates represent each of the principal construction disciplines, for example, architecture, engineering (civil, structural, mechanical and electrical) and project management. They may be graduates, fresh from university, or senior engineers with many years professional experience. In spanning this range the workshop offers a chance for the young to interact with the (too often remote) upper echelons, it brings different departments within AMEC into closer contact and it introduces collaborators from other companies to their colleagues in AMEC. This variety of encounter is recognized as being one of the keys to the success and popularity of the event.

## Outline of the workshop

The Designing Together workshops take place over a two-day period in a management training centre well away from the turmoil of the design office. This detachment from the office environment is seen as being critical to the success of the event, as the delegates are not available to work on projects or be contacted midway through the exercise. An experienced external facilitator, with a background in the construction industry, leads the workshops. The event is based around two key exercises: a preliminary exercise (the design of a newspaper), whose purpose is to accelerate the formation of a team mentality in readiness for the second, main, exercise, which is the design of a cladding system to be retro-fitted to an existing office building. Each of these exercises is punctuated by periods of discussion and reflection. These breaks enable the facilitator to introduce concepts, monitor the effectiveness of the team, and advise on possible improvements. The facilitator does not intrude his or her own comments in relation to the product itself, as this is not the purpose of the session. The facilitator's role is solely that of an enabler, prompting progress (process) and instigating negotiation and agreement between team members, these being the drivers behind early stage design activity (Steele, 2000).

## Icebreaking

Tuckman (1965) has observed that when teams are brought together for the first time they pass through four behavioural stages: 'forming', 'storming', 'norming' and 'performing', whereby team members progress from the awkward formality of first encounters, through clashes and argument, to reasonable levels of accommodation before finally establishing trust and giving mutual support.

If allowed to develop in its own time, this dynamic may take longer than a two-day workshop to move beyond the uncomfortable and unproductive forming stage. The Designing Together Workshop therefore accelerates this process by using the newspaper exercise as an 'icebreaker'. Each team is given access to paper, pens, pencils and a camera, and asked to report on recent news events in the local area. This typically leads to much discussion and banter as the teams conceive and deliver light-hearted news items. The major design element rests in generating the form and layout of the newspaper. However, much creativity is encouraged during the development of the stories, an activity that helps the attendees to break away from habitual preoccupations and behaviours. In addition, as the individuals begin to relax they tend to establish bonds with one another on a social level with the result that they begin to create the collaborative instincts and mechanisms which drive the remainder of the workshop.

## Main design exercise

The main workshop exercise involves the design of a cladding system. This problem was chosen because, in order to be solved effectively, it requires the input of each of the design disciplines. Teams are provided with a design brief, plans and sections of the building (including the dimensions of a typical office bay), and a copy of relevant technical guidance. In addition, they are provided with an extract from *How Designers Think* (Lawson, 1982). This is not intended to be used as a manual, but rather as stimulus material to reinforcing the idea that what is required of the participants is not simply a design solution but fresh thinking in relation to design itself. Once briefed, the teams are left to work on their own, with the only external interjection coming from the facilitator.

On the evening of the first day the designers have a relaxed dinner together. This provides a much-needed break from the toils of the design exercise and allows the attendees to exchange frank views in relation to the course itself and the effectiveness of their work to that point. It is often found that teams experience a marked improvement in their performance on the second day as a result of the air being cleared on a social, as well as professional, level.

As part of the evening's proceedings, a member of the AMEC senior management is asked to speak to the group and then take part in a debate concerning any subject the delegates wish to discuss. This invariably defaults to a discussion centring on design within the business, providing an opportunity for the participants to explore the management's views on interdisciplinary design and construction issues. This senior support has been commended by a large number of the delegates in feedback questionnaires, and is recognized as being of paramount importance to the continuing success of the course.

On completion of the exercise, on day two, the teams are given 30 minutes to present their outline proposals and to describe the design processes that they followed. A panel of three independent reviewers (senior managers from within AMEC) then offer critiques of the proposals with respect to the ease with which they might be developed into workable solutions that satisfy the brief. This section of the exercise allows the teams to get some independent feedback on the suitability of their proposals, and provides a much sought after opportunity for the designers to practise their presentation skills away from the client's gaze.

## Outcomes and benefits

The workshop approach to designing together was intended to achieve more than simply informing the delegates of good team working theory. What such a factually based course may deliver may very well be knowledge as a static commodity. Learning how to learn (and learning together) provides a mechanism for continuously gathering new knowledge and skills as a part of everyday life. AMEC recognizes that it is this core skill that enables a business to improve its performance over the long term.

## Feedback: success of workshop

Burdett (1999) has defined three measures of successful learning that might allow a course such as this to be assessed:

- evidence that those involved in the learning experience display the intended behaviour
- a determination of how the new ways impact on the performance of the organization
- a recognition that the change is sustained.

Though these markers may truly identify the success of a course, their assessment in practice is extraordinarily difficult. There are simply too many variables within the dynamics of design for the effect of a single training event to be assessed with certainty. To a degree, therefore, AMEC takes as a matter of faith that the course provides the intended benefits, but it is supported in this view by the delegates' responses as collected through formal feedback.

Since the Designing Together course has been run, five workshops, involving some 75 designers, have been held. Each delegate has completed a questionnaire to enable the training and development department to assess the suitability and popularity of the course. This auditing exercise, which is common practice within AMEC, allows training programmes to be developed to meet the employee's needs most effectively and indicates areas for further improvement. The feedback data from the Designing Together course (Figure 10.1) has shown it to be among the most effective and popular events within AMEC's training and development portfolio. This finding is further illustrated by

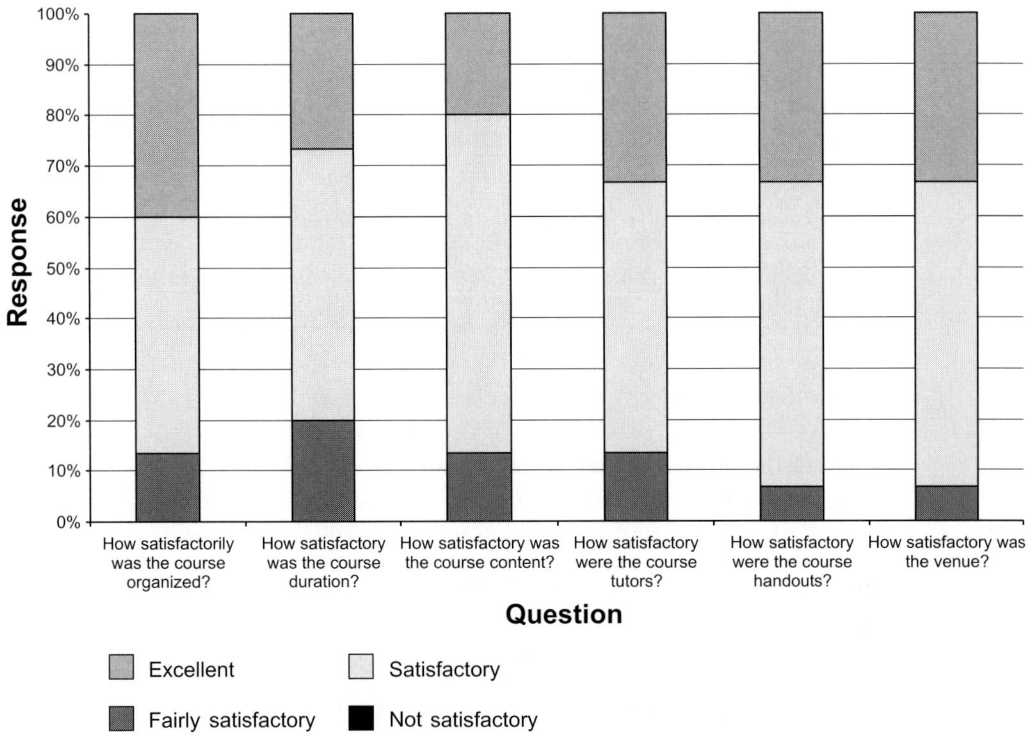


*Figure 10.1 Feedback data from Designing Together courses.*

the high level of demand from the company's designers, which has led to the course being now oversubscribed.

The qualitative responses submitted by attendees in the feedback questionnaires has also provided valuable insights into potential mechanisms for improving design practice. Many respondents have, for example, commented on the value of the facilitator during design sessions. By not having a stake in the design the facilitator was recognized as being able to contribute an unbiased influence over design process and decision making. Such facilitation, respondents believe, could lead to significant improvements to interdisciplinary working if imported into real-world projects. Consequently, members of the design management department are investigating this idea further.

Additional comments recorded as formal feedback by delegates are listed in Table 10.1.

As can be seen, the consistent positive feedback from the attendees reinforces AMEC's confidence that the Designing Together course provides real benefits.

*Table 10.1 Verbatim response data from questionnaire.*

| |
|---|
| *'the best course* [I have] *attended, all designers and managers should attend'* |
| *'there is a more to teamworking than meets the eye ...'* |
| '[The] *course tutors were good and* [the] *material was excellent ...'* |
| '[I] *will definitely try to apply this stuff in the office'* |
| '[I] *enjoyed the evening debate ... and the after dinner drinks* [!]' |
| *'I learnt about the other disciplines and myself (listening* [to team-mates] *does help!)'* |

## Feedback from the organization

The benefits afforded to AMEC through the introduction of the Designing Together workshops may be difficult to measure quantitatively, but there is considerable evidence that designers are more in tune with needs of their colleagues once they have attended the course. As long as the demand for the course persists, AMEC has committed itself to supporting the workshops and, on the recommendation of those who have attended, continuing to develop the content to reflect evolving needs.

Such has been its success that the format of the Designing Together course has been taken as a model for additional courses within AMEC. Currently being offered (or under development) are Building Together, Design Team Leader and Cost Planning Together. These are principally targeted at AMEC employees, but the company believes that benefits can be derived from including participants from further afield. In an industry where the formation of strategic alliances is becoming commonplace, it is apparent that collaborating organizations need to actively formulate strategies for effective joint working. There is little evidence to suggest that individuals or organizations can manage these relationships well if they rely solely on conventional education and habitual behaviour.

Furthermore, a Designing Together workshop based upon a live project, and involving clients, designers, constructors and suppliers, may offer many additional rewards. Used in this manner, the workshop can act as a 'rapid prototyping' session, where the project team can force themselves to deliver an initial design proposal in a constrained time span. Within such an exercise, multiple and rapid attempts at generating solutions become a means of developing a greater understanding of the problem. Evidence of the value of such events was offered by the designers involved, who suggested that the facilitated design exercise saved several weeks on the programme and was invaluable as a means of efficiently developing the brief. This approach parallels what the International Group for Lean Construction has termed 'scheme design in a day', which is now recognized as eradicating wasteful iteration from the early stages of the design process (Ballard, 2000).

## Research interests

AMEC's connection with the IDBE course at the University of Cambridge led to collaboration on a research project funded by the Engineering and Physical Sciences Research Council (EPSRC). The project, entitled 'Mapping the Design Process During the Conceptual Stage of Building Projects' (MDP), involved several construction industry organizations and used Designing Together workshops as a means of developing and subsequently field-testing a generic description of the phases and activities encountered during conceptual design. The intent was to develop a framework of actions to guide design teams during this early strategic design stage.

Two workshops were held during the course of the research project. The first of these was a Designing Together workshop involving designers from AMEC only. The second used the same format but included designers from a number of organizations with whom AMEC had recently been working or with whom they were due to work. These workshops not only enabled the framework to be developed and tested, but also enabled insights to be gained into the differences in working and social practice within teams whose members are, in the first case, familiar with one another, and in the second recently introduced. This project is described in greater detail in Chapter 14.

The researcher, the principal author of this chapter, rejoined the AMEC organization as part of the Innovation and Technology team on completion of the research project and has been developing the findings of the work for application in the company. He is also a facilitator on the Designing Together workshop courses and contributes to the teaching on the Cambridge IDBE course. The workshop, which began on the course, therefore crossed over into industry and spawned research which is now being fed back, providing a beneficial cross-fertilization between industry and university.

## Conclusions

Interdisciplinary teams offer the potential for innovative and high-quality solutions to building design problems. This owes much to the fact that effective multi-disciplinary teams develop synergistic approaches to their work by integrating and expanding the diverse perspectives brought by their individual members (Steele and Murray, 2000). These individual perspectives, which arise from the processes and outcome of interdisciplinary group working, can be both highly creative and innovative as the team benefits from an increased diversity in knowledge and understanding. However, it is important to recognize that a lack of awareness of, and sensitivity to, differences in disciplinary working practices and objectives can also have negative effects on the workgroup. For individuals from a variety of disciplinary backgrounds to work effectively as a team, each individual must not only have a specialist discipline, but must also be an effective communicator and negotiator, and must promote synergy between the disciplines and strive for shared understandings. It is these skills, and the ethos that their utilization induces, that has been delivered, in part, through the introduction of the Designing Together workshops.

## Acknowledgements

The authors would like to acknowledge the input of the AMEC organization, in particular AMEC Capital Projects Construction.

## References

Archer, L. B. (1984) Systematic method for designers. In *Developments in Design Methodology* (ed. N. Cross). J. Wiley & Sons, Chichester.

Austin, S., Steele, J., Macmillan, S., Kirby, P. and Spence, R. (1999) Using training workshops to map interdisciplinary team working. In *Proceedings of the Chartered Institute of Building Services Engineers (CIBSE) National Conference 1999*, Harrogate, UK.

Ballard, G. (2000) Positive versus negative iteration in design. In *Proceedings of the Seventh Annual Conference of the International Group on Lean Construction (IGLC)*, Brighton, UK.

Barnard, G. (1995) *Cross-cultural Communication: A Practical Guide*. Cassell, London.

Burdett, J. (1999) Leadership in change and the wisdom of the gentlemen. *International Journal of Participation and Empowerment*, **7**(1).

Buzan, T. (1995) *The Mind-Map Book*. BBC Books, London.

Cross, N. and Clayburn-Cross, A. (1996) Observations of teamwork and social processes in design. In *Analysing Design Activity* (eds. N. Cross, H. Christiaans and K. Dorst). J. Wiley & Sons, Chichester, pp. 291–317.

Frankenberger, E. and Auer, P. (1997) Standardised observation of teamwork in design. *Research in Engineering Design*, **9**, 1–9.

Lawson, B. (1980) *How Designers Think*. Butterworth Architecture, Oxford.

Macmillan, S., Steele, J., Austin, S., Kirby, P. and Spence, R. (2001) Development and verification of a framework for conceptual design, *Design Studies*, vol. 22, no. 2, 169–191.

Parker, A. and Steele, J. (1998) Improving the effectiveness of the conceptual design process by learning from other industries. In *Proceedings of the Chartered Institute of Building Services Engineers (CIBSE) Annual Conference*, Bournemouth, pp. 69–83.

Rittel, H. and Webber, M. (1973) Dilemmas in a general theory of planning. *Policy Sciences*, **4**, 155–167.

Steele, J. (2000) The interdisciplinary conceptual design of buildings. *Ph.D. Thesis*. Department of Civil and Building Engineering, Loughborough University.

Steele, J. and Murray, M. (2000) Constructing the team: a multi-cultural experience. In *Proceedings of the Chartered Institute of Building Services Engineers (CIBSE) National Conference*, September, Dublin.

Steele, J., Austin, S., Macmillan, S., Kirby, P. and Spence, R. (1999) Interdisciplinary interaction during concept design. *Proceedings of the 15th Annual Conference of the Association of Researchers in Construction Management 1999*, Liverpool, pp. 297–305.

Tuckman, B. W. (1965) Developmental sequence in small groups. *Psychological Bulletin*, **63**, 384–399.

# Part 3

# Design teams observed

# Introduction

## *Sebastian Macmillan*

As Part 1 of this book clearly shows, there is a strong desire within the construction industry and among its clients to improve the quality of the built environment, and a recognition that this can only be achieved through better working practices and relationships. With the support and encouragement of government, including all those client departments that are part of the Better Public Buildings Group, the construction sector is currently in the middle of a radical programme to re-engineer its business processes.

The industry is not being expected to achieve these changes unaided; a number of related mechanisms have been developed to collect and disseminate new knowledge to aid the industry. The Construction Best Practice programme was announced on 4 July 1997 and officially launched in November 1998. Funded by the DETR and steered jointly by the Department and the Construction Industry Board, it is intended to raise awareness of the benefits of best practice and provide guidance and advice to UK construction and client organizations so that they have the knowledge and skills required to implement change. Its main focus is transformation of outmoded management practices and business cultures. The Movement for Innovation (M4I) was established in 1998 to help the industry meet the Construction Task Force's objectives. Its mission is to lead radical improvement in the construction industry in value for money, profitability, reliability and respect for people, through the demonstration and dissemination of best practice and innovation.

DETR's Construction Research & Innovation Programme sponsors innovation and research in construction. Its aims are to help bring forward knowledge and longer term thinking in support of its strategy for sustainable construction and the 'rethinking construction' agenda. The Construction Research and Innovation Strategy Panel (CRISP) works with industry and government to help identify industry's strategic needs and set a research and innovation agenda. Construction research has also been supported by the Engineering and Physical Sciences Research Council in a specific programme – Construction as a Manufacturing Process – which was launched in 1995. This, together with two related LINK programmes funded jointly by DETR and EPSRC – Integration in Design and Construction and Meeting Clients' Needs through Standardization – have an emphasis on research into business processes with the aim of generating new knowledge to help the sector reduce costs, increase competitiveness, and improve quality, productivity and predictability. Additionally, the Foresight programme of the Department of Trade and Industry supports a Construction Associate Programme (within the Built Environment and Transport Panel) to help the industry identify and anticipate future trends.

As the introduction to this book showed, designers have a key role in the production of the built environment. Designer practitioners, like other players in the production of the built environment, are reviewing their business processes, as shown in Part 1. The education of designers is also changing, as shown in Part 2.

In this third part of the book, we address the contribution that research can make towards helping designers improve their design processes. The ideas reported here are based on the premise that a better understanding of interdisciplinary processes can help to improve communication and collaboration; and that better interdisciplinary collaboration will in turn help to ensure design outcomes fully reflect the combined expertise of the project participants.

In Chapter 11, Hawkes looks at the products of the design process and, by charting the historical evolution of a set of exemplary low-energy office buildings, demonstrates the changing relationships between fabric and plant. He shows how important historical awareness is to the development of new designs, and suggests that the evolutionary history of building types should be taught to all disciplines involved in the construction process. His historical survey leads him to conclude that the traditional distinctions between the professions should be progressively transformed to allow a richer and more creative collaboration.

In Chapter 12, Macmillan and French report on the early design development of a primary school in Essex. A professionally facilitated two-day interdisciplinary design workshop was held to bring together the legal client, user client, and the key disciplines involved in the design. The activities in the workshop are described, together with the responses of the participants, all of whom were positive about the experience. Drawing on this experience and on previous exercises both in Essex and elsewhere, the chapter offers research-based good practice guidance about how to organize and manage similar workshops.

Chapter 13 draws on the management science literature on teamwork and on the results of observing in detail three interdisciplinary design teams. It offers a definition of teamwork, and identifies the conditions necessary to ensure effective interdisciplinary teamwork during the design process. It discussed open communication, dealing with conflict, managing team dynamics and improving collaboration. It also suggests guidelines for running design team meetings. An extended list of references on teamwork is also given.

In Chapter 14, Steele and Macmillan discuss the results of a research project concerned with mapping the early stages of the design process. These stages can be vibrant and dynamic as a new team is formed, and they are crucial in terms of the fundamental decision that are taken about the form and fabric of the building. Yet often design teams have to reinvent the process every time they start a new project. By using case studies and reviewing the existing literature, the authors propose a 12-stage model for concept design intended to guide the process and promote collaborative design development. The chapter also reports on how the model was turned into a prototype interactive Internet-based system and tested on a live project.

CHAPTER ELEVEN

# Beyond the environmental tradition: lessons from the past for the architecture of the future

*Dean Hawkes*

## Summary

This chapter examines the proposition that knowledge of past designs plays an important role in the design of new buildings. Earlier work by the author on the idea of the *stereotype* as an aid to environmental design is reconsidered, and the environmental typology is extended to incorporate recent stages in the evolution of the office building. The chapter concludes with a discussion of the implications for interdisciplinary design practice and for the education of designers.

## Introduction

For many years, as teacher, researcher and practitioner, I have been preoccupied by the relationship between, in convenient, albeit imprecise, terminology, *art* and *science* in the creation of buildings. In the field of environmental design, where much of my work has been located, it is particularly important to be clear about this relationship in defining the relationship between the environmental function of the form and fabric of a building – the very material of the *art* of architecture – and the role of mechanical systems of environmental control – which are, in most manifestations, primarily questions of technology or *science*. In both my academic and professional work I have found that these questions are frequently illuminated and informed by reference to past instances. In other words that the production, perhaps even the *invention*, of solutions to present problems benefits from historical understanding.

My aim in this chapter is to put this hypothesis to the test. I will try to do this by proposing an evolutionary model of design – a kind of architectural Darwinism. This is a theme which I first explored over twenty years ago in my essay 'Types, norms and habit in environmental design' (Hawkes, 1976), and I welcome the opportunity to revisit and test the argument in the light of the experience of the intervening years.

## The environmental tradition

In 1995 I published a collection of essays on themes in the theory and practice of environmental design under the title of *The Environmental Tradition* (Hawkes, 1995a). The Foreword to the book was written by Robert Maxwell, who suggested that the essays, 'never oversimplify in the attempt to assert a general theory of scientific design' and that they are, 'conscious of the loose fit, that architects know only too well, between form and performance: a space in which cultural pressures can produce strange distortions'.

These observations, in my view, identify something essential in defining and understanding the nature of architecture. First, that it is not susceptible to reductivist, 'scientific' theories of design, and, second, that the ostensibly predictable, technically based aspects of architecture – such as environmental design – are conditioned as much by 'cultural pressures', in the broadest meaning of the term, as by logical analysis and precise calculation. It was precisely this question which was addressed 30 years ago by Alan Colquhoun (1967) in his seminal essay 'Typology and design method', in which he wrote, 'In spite of the development of the scientific method, we must still attribute ... social or iconic values to the products of technology and recognize that they play an essential role in the generation and development of the physical tools of our environment'. This essay was a key influence on in the development of my argument for the existence and use of 'stereotypes' in design, which was proposed in 'Types, norms and habit'. There, referring specifically to the problem of the design of office buildings in the early 1970s, I sketched an 'Evolutionary Tale' in which I traced the development of the British office building from the turn of the century through the description of six distinct stereotypes (Figure 11.1).

The operational argument for the stereotype is that, '... it is simply a generally held notion about the nature of a good solution to any recurrent design problem ...' which plays '... a creative role by allowing the designer to begin ... from a reasonably confident position'. In addition, the definition and characterization of these particular office building stereotypes allowed me to make an observation about the transformation of environmental design in the twentieth century which they reveal. This is most strongly illustrated by the comparison of the first

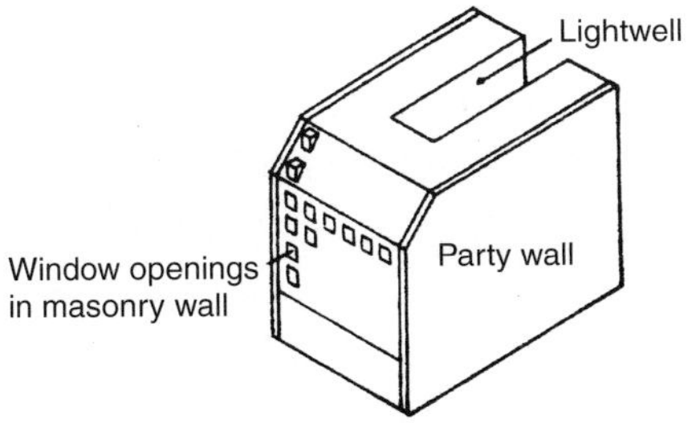


3.1
Stereotype 1, *c.* 1900: London Building Acts, lightwell form.

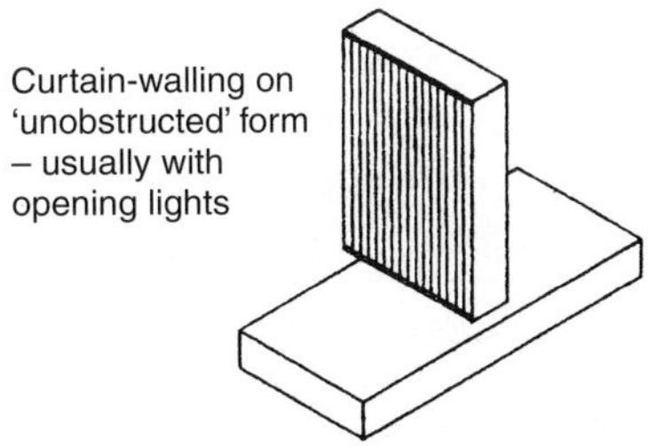


3.4
Stereotype 4, *c.* 1955: slab-on-podium.

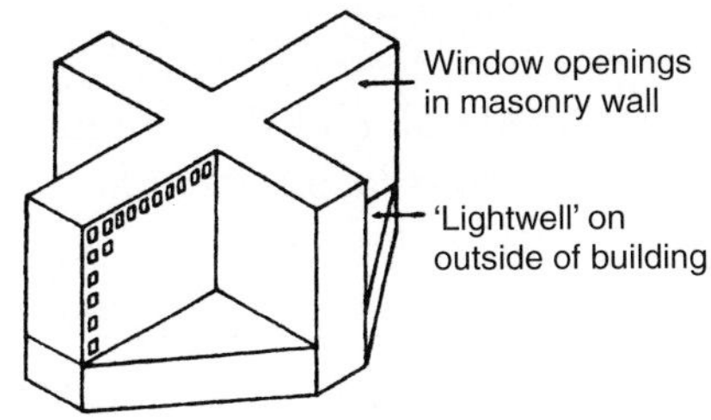


3.2
Stereotype 2, *c.* 1930: 'lightwells on the outside'.

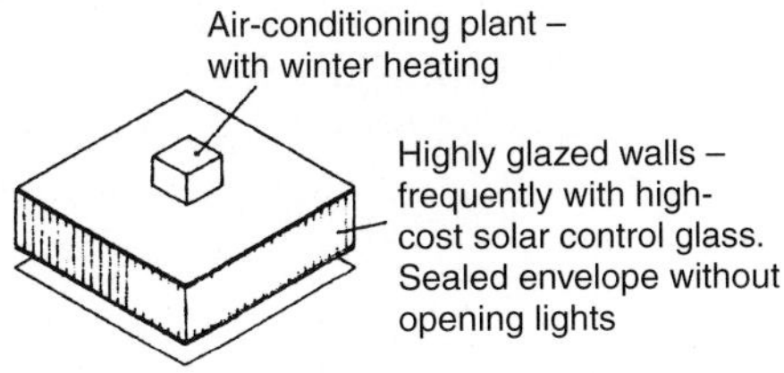


3.5
Stereotype 5, *c.* 1970: deep-plan, air-conditioned.

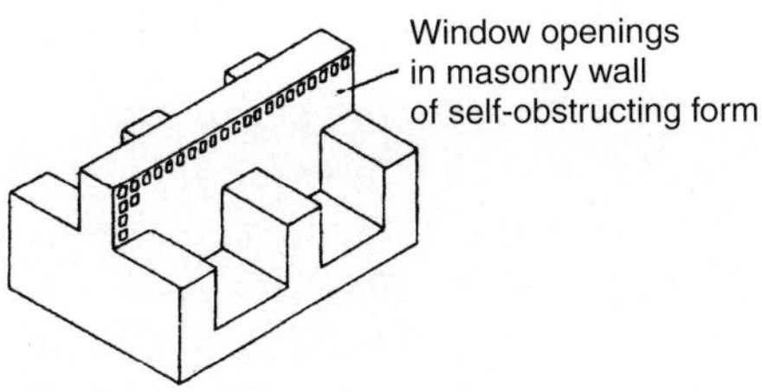


3.3
Stereotype 3, *c.* 1945: postwar daylighting standards.

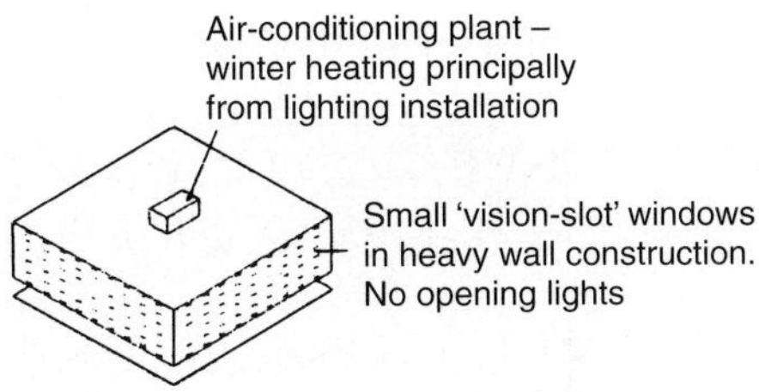


3.6
Stereotype 6, *c.* 1975: Integrated Environmental Design (IED).

*Figure 11.1 Office building stereotypes.*

and last types defined. In the first the relatively loose relationships between its components – plan form, floor-to-ceiling heights, window sizes, the type of mechanical plant, and the broad notion of comfort of the time – permitted a free adaptation of its broad characteristics in the design of any specific building. On the other hand, the last type – based on the 1970s 'theory' of 'Integrated Environmental Design' (IED) – rests upon a 'scientifically' determined set of relationships between, form, construction, plant and 'comfort' which, because of its closed logic, prohibits inventive elaboration. But, particularly with the benefit of hindsight, the most useful function of the 'stereotype' lies in:

> *... the realization that the latest stereotype does not supersede all others. There is, in fact, a store of accumulated experience which contains* **all** *[new emphasis] previous solutions and which will be enlarged in the future with the addition of new examples informed by changing building technology, organizational ideas and physical, social and cultural contexts. This view demands a return to earlier stereotypes to see what they offer as potential solutions to present-day problems ... The search should be for the most appropriate solution ... the first step would be to examine the store of stereotypes rather than simply to accept a single current notion .... The result would be a richness of solutions inspired by the particular nature of each problem and the achievement of building science would find its true place within a proper understanding of the nature of design.*

There is a resemblance between this argument and the proposition made in the early years of the century by Lethaby (1912), who in his little book *Architecture: an Introduction to the History and Theory of the Art of Building*, wrote:

> *Modern builders need a classification of architectural factors irrespective of time and country, a classification by essential variation. Some day we will get a morphology of the art by some architectural Linnaeus or Darwin, who will start from the simple cell and relate it to the most complex structures .... So long as the whole field of past architectural experiment is presented to us accidentally only under historical schedules, designing architecture is likely to be conceived as scholarship rather than as the adaptation of its accumulated powers to immediate needs.*

Lethaby's work was brought to my attention by Philip Steadman, who has written authoritatively on evolution in architecture (Steadman, 1979). At the turn of the century, Lethaby's concern was to rescue architectural history from its function as a source of stylistic reference and to render it of value to modern design practice – as understood in Britain at that time. His own very few buildings hint at the kind of synthesis between history and contemporary needs which he advocated. This is, perhaps, most clearly demonstrated in his exquisite church of All Saints at Brockhampton in Herefordshire (1902). Behind an appearance which, particularly with the maturation of nearly a century, seems to

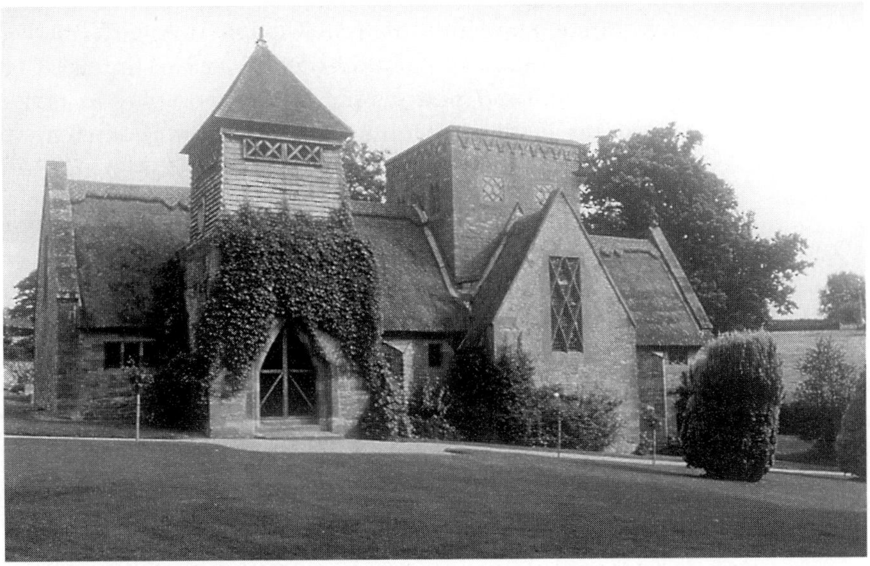

*Figure 11.2 All Saints, Brockhampton, 1902, W. R. Lethaby.*

have grown organically out of its site (Figure 11.2), we find a roof structure of *in situ* concrete and a warm air heating system which is fully integrated into the fabric of the floor and walls.

The emergence of the Modern Movement, with its rejection of history, shifted the theoretical ground towards the idea that designs could, and should, be based upon the 'scientific' understanding of 'first principles', as Summerson (1957) explained in his essay, 'The case for a theory of modern architecture'. This made a project such as Lethaby's redundant. But that 'hard-line' modernism has long been under critical scrutiny, as is demonstrated by, for example, Colquhoun's arguments in *Typology and Design Method*. The case for a wholly 'objective', 'systematic' basis for the production of architecture no longer has credence in the theoretical discourse. We may, therefore, look again to the possibility of history, properly understood and analyzed, playing a part in the production of inventive solutions to the problems which confront contemporary practice.

At this point I should stress that this is not an argument for nostalgic revivalism. That was not the point of Lethaby's project either. The intention is to make the data contained in the historical record available to the contemporary practitioner in a way which helps to develop designs that combine effective technical performance with an appreciation of the 'cultural pressures' which bear crucially upon all architecture of significance. I am mindful of the wisdom in Robert Maxwell's strictures about the dangers of 'general theory' and of the need for 'loose fit ... between form and performance'. The best way to make the point is through the medium of buildings. It is in the concrete reality of designs that the abstractions of theory are best illustrated and validated.

## Case studies

My original set of stereotypes pre-dated the emergence of the atrium office building. Since the late 1970s this has become one of the most clear-cut and effective forms for office design – a potent demonstration of the value of the stereotype. In 1978 Richard MacCormac and I published a short article in the *RIBA Journal* (Hawkes and MacCormac, 1978) in which we developed the fundamental typological distinction of *courts* and *pavilions* which Leslie Martin and Lionel March first articulated in the 1960s (Martin and March, 1966). By adding a glazed roof to the court form we demonstrated that it would be possible to achieve a configuration of office space which would be comfortable and energy efficient in the British climate without the need for air-conditioning. In essence, the glazed courtyard combines the advantages of the traditional narrow plan office building with those of the deep plan type, while avoiding their disadvantages. In addition to this development of Martin and March's theoretical studies, we were equally influenced by specific buildings, such as Frank Lloyd Wright's Larkin Building in Buffalo and Alvar Aalto's Rautatalto Office Building in Helsinki, both of which have top-lit central spaces. In other words we were searching the store of precedent, exactly in the manner proposed in 'Types, norms and habit'.

At about the same time, Arup Associates were working on their design for Gateway Two at Basingstoke – proof of the *zeitgeist*? – (Figure 11.3). This is widely regarded as the seminal atrium office building in Britain. Unlike many of its successors this is daylit and naturally ventilated, with air-conditioning restricted to specialized local areas (Hawkes, 1983). The project architect at Gateway Two was Rab Bennetts, who now practises as Bennetts Associates and has continued to develop and refine the Gateway theme in subsequent buildings. The Powergen Headquarters Building at Coventry (Figure 11.4) is a subtle typological transformation of Gateway in which linear, daylit, naturally ventilated office spaces flank a long atrium (Hawkes, 1995b). The office spaces are open to the atrium and the facades reflect the difference between their northerly and southerly orientations in their detailed design.

The point in illustrating the Powergen Building is to show how the stereotype serves to inform the production of a design which responds

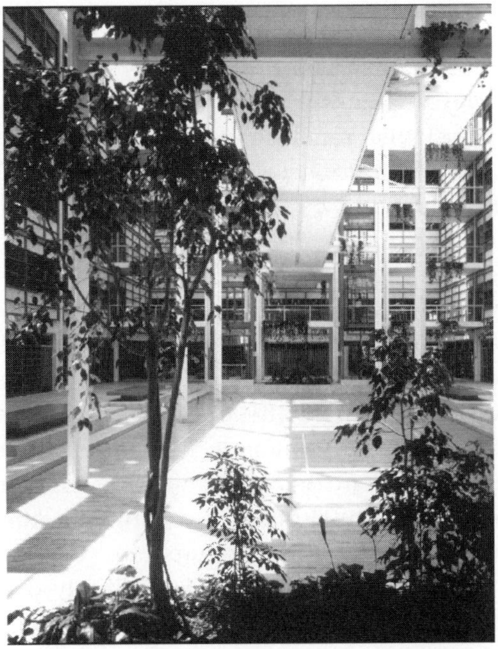

*Figure 11.3 Gateway Two, Basingstoke, Arup Associates.*

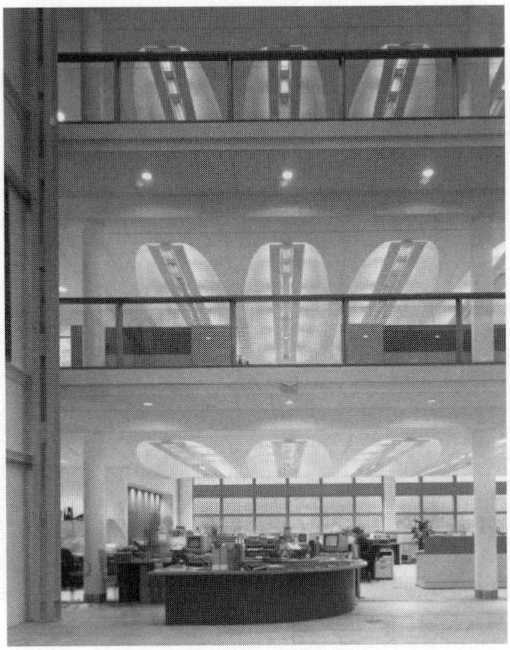

*Figure 11.4 Powergen Headquarters, Coventry, Bennetts Associates.*

to the specific conditions of programme and site. The building is securely based on the authority of the type, and on the technical and social attributes which it possesses, but it is a creative reinterpretation, not a slavish repetition.

The design of comfortable, energy-saving office buildings may be approached in a number of ways. Feilden Clegg's recently completed Building 16 at the Building Research Establishment (Figure 11.5) demonstrates the continuing validity of the traditional narrow plan office, as it is transformed by the superimposition of the devices of passive solar design (Bordass *et al.*, 1997). Again the distinction between north and south is clearly expressed in the design of the facades, but here the traditional elements are supplemented by the expressed ventilation stacks and light shelves. In terms of the stereotype argument this building shows how new knowledge and components may transform an existing type, investing it with a new level of performance. In this, perhaps, we can identify the emergence of a new stereotype. That is, indeed, explicitly the intention in BRE's requirement that the design should be replicable.

My final case study is the Friary Project at Maldon in Essex (Figure 11.6) designed by my own practice of Greenberg and Hawkes (Colbourne, 1994). This project consists of two buildings in the historic centre of Maldon and was conceived within all of the constraints which

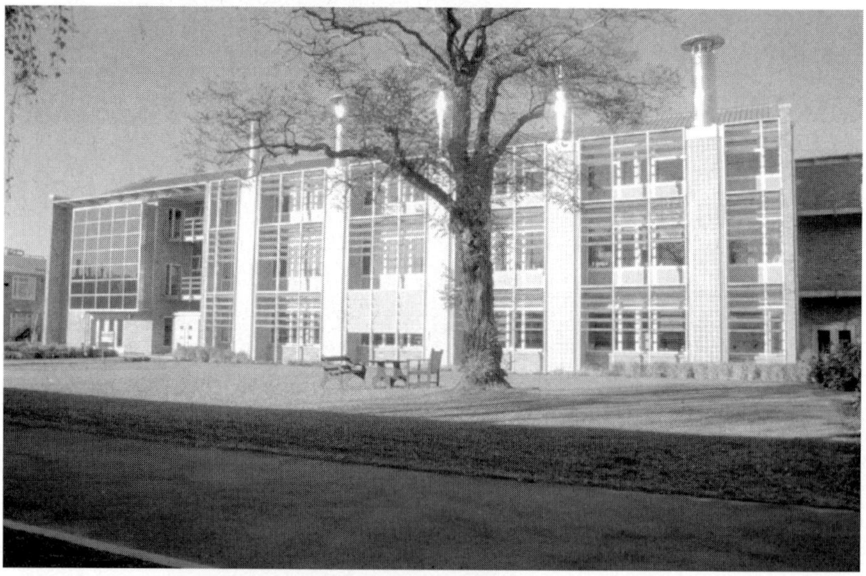

*Figure 11.5 Building 16, Building Research Establishment, Garston, Feilden Clegg Architects.*

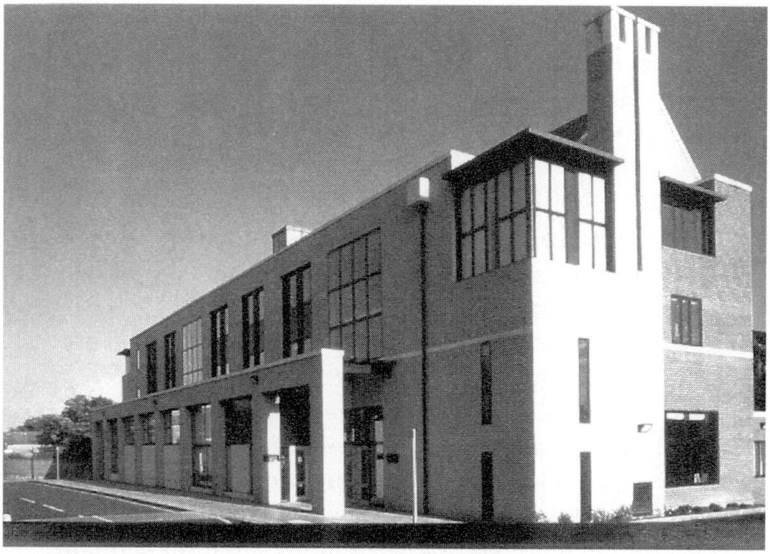

*Figure 11.6 The Friary, Maldon, seen from the north-west, Greenberg & Hawkes.*

are typically imposed in such situations. One of the buildings houses a branch of the County Library and offices for the County Social Services Department. The design of the offices responds to the specific structure of the department, with a mix of cellular and open plan spaces. In its fundamentals the building is a traditional narrow plan type. In comparison with both Powergen and Building 16 the facade design is much more conventional, with regularly spaced window openings and a complete absence of external devices. The principles which underlie the building are those which are codified in the *LT Method* of low-energy design, developed in Cambridge by Nick Baker and Koen Steemers (1997). The aim is to establish the dimensions of windows, in relation to the size and orientation of the spaces they serve, to optimize the balance of energy used for lighting and space heating. In this building there is a clear difference between the small windows on the north facade, which serve the cellular offices, and the larger openings to the south-facing open plan offices. Incidentally, the LT Method was not directly applied in the design process of the building, but its principles were sufficiently understood through many years of academic association with Baker and Steemers for them fundamentally to inform the design (Hawkes and Steemers, 1991). In comparison with Building 16, this design represents a less radical transformation of the narrow plan stereotype, but is, nonetheless, effective in achieving its environmental goals. In terms of biological evolution it is, perhaps, close to the subtle variations which Darwin detected in the beaks of his finches. But such small variations are as valid in architectural evolution as they are in nature in achieving an effective response to a changing environment.

## Conclusion

In conclusion I want to try to make a connection between theory and practice and to offer some thoughts on the future of education for the design professions.

My original paper on the idea of the 'stereotype' was written just when the objectives and methods of environmental design were being challenged by what we then called the 'energy crisis'. At that time the almost unchallenged assumption was that the environmental functions of buildings would increasingly be met by mechanical plant, the realm of the engineer, rather than by the form and fabric of the building, the traditional territory of the architect. It was also assumed that this shift would redefine the nature of the relationship between the members of the design team. In the present context it is not insignificant that the last 'stereotype' in my original typology was based upon the principles of 'Integrated Environmental Design'. Now, after two decades of further evolution of the office building type, that assumption is no longer valid. Increasingly we see, as my case studies show, that the primary environmental parameters are often again addressed by the *architecture* of the building, and that the relationship of fabric and plant is, in such buildings, fundamentally redefined. This must have deep implications for the way in which the design team is constructed and functions. As I see it, the need remains, is actually reinforced, for a philosophy of integrated, or perhaps more appropriately, *interdisciplinary*, design. But this now is, or should be, one in which the traditional distinctions between the professions are progressively dissolved to allow a richer and more creative collaboration. This has profound implications for the professions and for the education of their members.

We are no longer conditioned by the strict dogmas of the Modern Movement. The theory and practice of architecture are characterized by a more inclusive view. The intellectual necessity, felt by the 'pioneers', to escape from the historicist confusions of the nineteenth century has served its purpose, and now the value of historical knowledge is almost universally acknowledged. The 'discipline' of architecture as an academic enterprise is founded upon its history. But for history to be of more than just generalized value – a kind of colour supplement – it must be brought to bear upon all aspects of contemporary practice, not just those which have traditionally been the preserve of the architect. I propose that, just as there is an obligation in architectural education to convey a substantial grounding in building science and technology, so there should be a parallel obligation for the development of a historical sensibility in the education of all members of the design team.

In 'Typology and design method', Colquhoun wrote:

> *In spite of the development of the scientific method, we must still attribute ... social or iconic values to the products of technology and recognize that they play an essential role in the generation and development of the physical tools of our environment .... This fact is concealed from us because the intentions of the design process are 'hidden' in the overt details of the performance specifications.*

What I have tried to show in this chapter is that it is possible to structure the data of the past in a way which makes it relevant to the problems of present-day environmental design. By identifying distinct *types* of solution these become available for use as the starting point of new designs. In the hands of inventive designers the *stereotype* provides a springboard for invention, and thus leads to the enrichment of the database, creating new types for future use. But the idea of *type* also allows these essentially technical aspects of design to be connected to the wider cultural context, and thus helps in reconciling the relationship between *art* and *science*, achieving true integration of design in the built environment.

## References

Baker, N. and Steemers, K. (1997) *Energy and Environment in Architecture*. E. & F. N. Spon, London.

Bordass, W., Hawkes, D. and Roaf, S. (1997) Building 16, Building Research Establishment, Garston. *RIBA Journal*, April.

Colbourne, C. (1994) Building study: The Friary Project, Maldon, Essex. *The Architects' Journal*, 15 September.

Colquhoun, A. (1967) Typology and design method. First published in *Arena*, **83**, June. Reprinted in Alan Colquhoun (1981) *Essays in Architectural Criticism: Modern Architecture and Historical Change*. MIT Press, Cambridge, MA.

Hawkes, D. (1976) Types, norms and habit in environmental design. First published in *The Architecture of Form* (ed. L. March). Cambridge University Press, Cambridge.

Hawkes, D. (1983) Building study: Gateway Two, the Wiggins Teape Building, Basingstoke. *The Architects' Journal*, August.

Hawkes, D. (1995a) *The Environmental Tradition: Studies in the Architecture of Environment*. E. & F. N. Spon, London.

Hawkes, D. (1995b) 'Building study: Powergen Headquarters, Coventry. *The Architects' Journal*, 2 March.

Hawkes, D. and MacCormac, R. (1978) Office form and energy use. *RIBA Journal*, June.

Hawkes, D. and Steemers, K. (1991) Research into practice: a case study of the application of technical studies in architectural design. *Renewable Energy*, **1**(3/4).

Lethaby, W. R. (1912) *Architecture: an Introduction to the History and Theory of the Art of Building*. Williams and Northgate, London.

Martin, L. and March, L. (1966) Built form and land use. *Cambridge Research*, April.

Steadman, P. (1979) *The Evolution of Designs: Biological Analogy in Architecture and the Applied Arts*. Cambridge University Press, Cambridge.

Summerson, J. (1957) The case for a theory of modern architecture. *RIBA Journal*, June. Reprinted in Summerson, J. (1990) *The Unromantic Castle*, Thames and Hudson, London.

**CHAPTER TWELVE**

# Successful teamwork from day one: running an interdisciplinary design workshop

*Sebastian Macmillan and Christopher French*

## Overview

Buildings can be thought of as complex assemblies of a wide variety of products, brought together to house human activities and meet the demands and expectations of occupants – in terms of comfort, safety, appearance and value. The number of disciplines required to design a building is already wide, and probably increasing as the range and functionality of building components expands. The early stages of the design process are all-important, for key decisions taken then influence the subsequent development of the design and how well it meets or, potentially, exceeds expectations. How can the expertise of a wide range of specialists in separate disciplines, who may not have worked together as a team before, be fully used at these early stages? This chapter describes the use of interdisciplinary design workshops at the early stages of the design process, and suggests how to organize and manage them. A detailed example is also given. Like Chapter 13, the recommendations arise from research at Cambridge University's Martin Centre for Architectural and Urban Studies. The intention is to provide a research-based digest of good practice in the running of interdisciplinary workshops.

## Interdisciplinary design

The knowledge needed to design a building of any size and significance is spread across a large number of specialists representing different disciplines. Often too, 'the client' is a complex body made up of those who pay for the building, those who will occupy it, and those who will manage it. Client requirements are not only many and varied but may even be conflicting. Successful design can be understood as a process of negotiation between the varied requirements of 'the client' and the ideas and proposals made by designers until, through an iterative process, a solution that meets the requirements and aspirations of both the client body and the designers is reached.

Designers from various backgrounds regularly say that they would like to be involved earlier in the design process than they were in practice. This has been found consistently among young professionals on the Cambridge IDBE course, for example. They report being consulted only after key decisions have been taken – leaving them to react to others' proposals as best they can, instead of contributing their own expertise at the appropriate time.

The aim of interdisciplinary design teamwork is to give each building discipline, as well as 'the client', the opportunity to contribute its own knowledge at the appropriate design phase, and particularly at the early stages when key strategic design decisions are being made. At best, interdisciplinary design helps to harness the collective energy and expertise of all the members of the team so that designs are optimized in terms of efficiency, economy and quality. It has the additional benefit of creating a sense of common ownership of the design problem across the whole design team and the client body, and therefore a deep commitment to making the design work.

## Design team workshops

More often than not, the members of building design teams are based in different offices and drawn from different consulting firms. They may not have worked together before. Design workshops are intended to bring together in one place at the start of a project the key people who will fund, design, occupy and manage the project, with the aim of encouraging collaboration and communication between team members. Such events are known variously as intensive design exercises, co-location workshops or team-building exercises. They are intended to do far more than help the team members 'put a face to the name' next time they telephone each other. Rather, they help to launch a design project with a positive and enjoyable experience for the team members. They create an opportunity for team members to hear and appreciate each other's goals, and to identify common goals across the whole team. They help to engender a spirit of cooperation and collaboration between team members, leading them to trust one another's judgement. Perhaps most importantly, they encourage a vital sense of ownership of the design problem and of the emerging design solution across the whole team, in which responsibilities are shared by, not divided among, the team. Organizations that have used such workshops and had them written up include Essex County Council (Batty *et al.*, 1996), BAA and AMEC (Austin *et al.*, 1999; see also Chapter 10).

If one or more client representatives are present, this allows them to feel a part of the team – equally it gives all team members access to the client. The team hears directly from the client not just the bald design requirements, but about their experiences of past buildings and their aspirations and expectations for the new building – not just what they have, but what they want. It also enables the whole team to test its preconceptions with the client.

Often clients welcome the opportunity to hear the team's discussions and debates. A workshop removes some of the mystique commonly associated with the design process. Through participating in the decision-making process in this way, clients come to trust (and to be trusted by) the team. They too develop that vital sense of ownership of, and involvement in, the design outcome which helps to ensure they do their utmost to manage the facility effectively.

## Planning a design workshop

Forward planning of a design workshop is essential and should include the objectives, an event programme, and the domestic arrangements. The location is important – a neutral venue such as a conference centre is usually preferable, one that is not owned by the client or any of the design firms involved. This ensures that the usual office distractions are removed so that everyone can concentrate on the task at hand. It also helps to ensure that everyone meets on equal terms and do not feel they are visiting someone else's home territory. The location and ambience of the premises are also important for morale. The work can be intensive and stressful, and an attractive environment in a calm location will improve individual motivation as well as contributing to the feeling of a sense of occasion. Casual clothes may be encouraged. The workshop will need access to the usual equipment for such an event – a whiteboard, overhead projector and so on.

The objectives for the event – setting out what it is intended to achieve – together with a timetable need to be pre-planned. Typically this will be undertaken by a small workshop planning group of two or three key players who have experience of how much can be expected and how much achieved. It may be, for example, that the objectives for a workshop are to reach a certain stage of decision making. Depending on their training and personality, some participants may feel the need to produce a building plan, but, where the aim is to make key strategic decisions, it is important not to let the workshop be side-tracked by excessive concern for detail at this stage. Clearly the extent to which the scheme progresses towards the detailed design stage will vary according to the type of project and the time available.

A suggested programme for a two-day event is given in Box 12.1.

## Box 12.1 Typical outline programme for a two-day design team workshop

### Day 1

*Morning*

Round-table introductions

Purpose of the workshop

Discussion of workshop objectives and what should be achieved at the end of the workshop

Introduction to the project

- a briefing from the legal client or funding body
- a briefing from the user client or occupant representative
- presentation of state of the art in design of the building type (external or in-house speaker).

Development by the team of a set of key issues

*Afternoon*

Analysis of needs and circumstances

- the site and surroundings – constraints and opportunities
- internal requirements – characteristics required of each of the main spaces.

A search for solutions

- breakout groups to consider critical areas of the site and the building.

Plenary session with reporting back findings from the breakout groups

### Day 2

*Morning*

Review of the previous day's achievements

- review and evaluation of the options
- selection of limited number of options or themes for further exploration (possibly in groups).

Breakout groups (either the same as before or different constituted) make further explorations

*Afternoon*

Presentations from the groups and round-table comparison of each group's findings

Drawing together of the conclusions from the workshop

Identification of critical issues, further information, and any technical research needed

Round table discussion of what next – the follow-up actions needed by individual participants and others

Agreement on a programme and timetable for the next stage

Preparation of archive material for formal summary of the workshop

Presentation to outside parties (if required)

Review and evaluation of the workshop – advantages and disadvantages of this form of working, and how to take it forward and continue the dialogue

Workshop ends

## Selection and briefing of workshop participants

Selection of the participants is critical. Clearly members of the team need to be chosen for their expertise and experience. But some are better suited to teamwork than others, and positive outcomes are most likely to be obtained when participants have a willingness and ability to collaborate with others – people who are open to new ideas, who believe that 'two heads are better than one' and who do not feel threatened that others will tread on their professional patch. This demands people who are confident in their own abilities but who also respect the capabilities of others. Such people are most likely to share their knowledge willingly and be open to the complementary knowledge of others.

Less suited to this type of workshop are people who have what may be termed an over-familiarity with traditional roles and deep-rooted professional prejudices. Those, for example, who believe that the client should set out all the requirements in a comprehensive written brief, that the architect should prepare a sketch plan that responds to this brief, and that engineers should then be brought in to make the building stand up and be comfortable, are unlikely to work well in this kind of interdisciplinary setting.

Ideally the team will comprise representatives from:

- the legal or paying client, for example in the case of a local authority from the property services department
- the eventual users, for example, a school head-teacher, facilities manager, or general manager
- the design team, comprising, for example, a team leader, architect, quantity surveyor, structural engineer, mechanical and electrical services engineer, interior designer and landscape architect.

Participants will need to be briefed beforehand about the workshop's objectives and reminded to bring with them any reference sources that they will need access to, including examples of past designs. Participants may arrive at a workshop with preconceptions, but they should be discouraged from bringing preconceived design solutions.

## Specialist presentations

A key advantage of bringing the design team together at the beginning of a project is not only that they can be briefed as a team but also that specialist presentations can be given to them. A presentation may be from an acknowledged expert in a particular field (for example in a certain building type), or by one of the team researching and preparing a talk (for example about window design, green issues or intelligent buildings). One or more introductory talks can help to set a context for the building type. Having the team present for such a talk gives all the members the opportunity to discuss the issues with the speaker and among themselves.

It may be appropriate for the team to receive presentations by one or more of the team members – such as experience arising from earlier projects of a similar nature, or about the requirements for the present scheme. In the case of a school design, an example might be a summary of a typical day in a school by the head teacher. Where presentations contain lists, for example lists of desirables for a school, copies need to be available to team members; detailed verbal descriptions are unlikely to be recalled comprehensively. In including formal presentations, it is important that the workshop does not turn into an event where the participants feel they have simply been talked at most of the time.

## Setting workshop objectives

Broad objectives for the workshop – specifically, the stage that it is intended the proposals will have reached by the end of the workshop – need to be set by the workshop planning group. It may be that the workshop is largely convened to establish broad principles and strategies for the project, even if a detailed technical brief for the building exists. In this case the focus would be on, for example:

- arrangements for vehicular and pedestrian access,
- the location of the building and ancillary accommodation on the site,
- the footprint, form, orientation and general massing,
- the number of floors and maximum depth of occupied spaces, and
- broad principles for the fabric, windows, and environmental performance.

The advantage of focusing at this strategic level is to prevent designers focusing on the detail and, in their concern to deliver a complete sketch scheme, missing the bigger picture. It can also be valuable to discourage the team from feeling constrained by the detailed requirements of the brief.

Alternatively, the exercise may be to go further towards the development of a sketch plan, sections and elevations at say 1:100, complete with strategies for structural arrangement and servicing, and some preliminary costings.

The decision on how far to take the scheme will affect who should attend the workshop sessions. The client's presence is highly desirable for the early strategic decision-making stage. But it can also be useful later, even while the design team is engaged in 'drawing up'. The design will continue to evolve even during these later stages, and there are advantages in the client representatives being present, not least to maintain their sense of ownership of the project. This is preferable to their attending initially to contribute to the initial brief, then returning at the end of the workshop for a presentation of the final outcome. This latter process is likely to leave them mystified about what has happened in between and, potentially, unsure how the eventual outcome has been achieved.

There can be tension between the team leader's desire to deliver something quickly – a sketch plan, for example – and a desire by others in the team to debate fully options and alternatives. A carefully set programme can help resolve this tension. Typically, if the team can first agree shared values and objectives, actually producing the design may be straightforward.

## The role of facilitator

Highly interactive events which are well managed and in which everyone present is encouraged to participate fully can be a valuable start to a project – and a facilitator can be crucial in getting the best from such an event. The facilitator's role is to help the exercise achieve its objectives –

partly in terms of timekeeping and the delivery of the required outcome by the end of the event, but also by ensuring that all the delegates participate in the process and do not just feel they have been passive observers. He or she will carefully manage the event to give the opportunity for all delegates to offer their individual points of view and be heard, and also gain their commitment to a shared vision for the project.

Team leadership is a separate function from workshop facilitator. The role of team leader is to ensure that the emerging design satisfies the needs of the client, and that issues and ideas raised by team members are incorporated into the team's decisions. A rapporteur who keeps notes of the progress of the event and records ideas and decisions on the whiteboard is also useful at a workshop. The team leader is likely to be too busy to double up as rapporteur.

There are advantages in selecting an independent professional facilitator who is not associated directly with any of the participants, but who is skilled in getting the best from workshop teams. Their independence will help to ensure they do not show favouritism to any of the participants and treat all contributions to the workshop on their merits. They may also be more determined to hold the team to an agreed programme and ensure the workshop meets its objectives. However, just as there are advantages in the workshop members already being familiar with one another, so it can be valuable if the facilitator has an established working relationship with them.

Ideally the facilitator will be:

- knowledgeable about the building type being designed,
- knowledgeable about the process of design and the set of activities that the team needs to undertake,
- interdisciplinary in outlook, uncommitted to an individual discipline, and
- capable of understanding the varied terminology of clients, facilities managers, architects, quantity surveyors, engineers, project managers, landscape architects and other team members.

Perhaps the key personal quality required of a facilitator is to be a good communicator, someone who can introduce humour to diffuse conflict, and who is capable of inspiring the team by maintaining a dialogue and producing ideas on those occasions when the team has run out of them.

The facilitator has a number of administrative and technical tasks:

- to introduce team members and set a context for the exercise, including discussing the issues involved in achieving effective teamwork

- to set a timetable for presentations, plenary sessions, group working sessions, and periodic progress reviews, and agree this with the team members
- to run the exercise in accordance with the timetable
- to focus on enabling the team to meet its objectives
- to 'chair' progress reviews and feedback sessions following individual group sessions, including ensuring that all team members are given the opportunity to make a contribution
- to encourage the team to recognize that ideas can be brought forward by anyone, as they can trigger something that others can usefully build on.

It can be valuable to maintain a flexible agenda for a workshop to allow the facilitator to vary how the exercise is run according to the progress made by the team.

## Suggested ground rules

Ground rules for the design workshop depend a great deal on how well the participants already know one another, the extent to which they have worked together before (if at all), and the arrangement of the event itself. Encouraging participants to be punctual will help the meeting keep to time and meet its objectives. On the other hand, some flexibility may be in order – comfort breaks may be a time when valuable exchanges of views take place, albeit informally, which will be lost if the timetable is interpreted too rigidly. Design workshops are not like business meetings where information is transacted according to a fixed agenda. If the team members are not familiar with one another, it may be worth asking them to:

- be open and frank
- address issues not personalities
- be positive and constructive
- keep an open mind
- try to achieve consensus.

Other ground rules to explain to the team are:

1. preconceptions in design are essential – but they should be relinquished once they are shown to be obsolete
2. everyone should be willing to put their own ideas on the table – not feel constrained and compromise at the first sessions. Ideas from one person can trigger things that others can build upon
3. good guesses are often valuable as initial responses and should be explained to the design team members

4. teams should be tolerant – there should be no recriminations when earlier guesses are checked and, if necessary, revised.

In order to get participants who have not worked together before to know one another quicker, they may be asked to introduce themselves briefly by for example, describing their hobby, or stating which famous personality or historical character they would most like to meet and why. A simple management game, perhaps played between two teams, can also help to break the ice.

## Breakout sessions

If the design team is too large to make progress as a single group, or if there is a need for working in parallel to make faster progress, breakout groups may be appropriate. Separate breakout groups might, for example, explore the issues surrounding external circulation, contours and landscape, or building shape and block form, in more detail. A rapporteur from each breakout group needs to present the results of their deliberations to the whole team.

### Box 12.2 The experience of Essex County Council in running interdisciplinary design workshops

Essex County Council has been committed to collaborative teamwork at the early stages of design for a number of years specifically for their school building programme (Batty *et al.*, 1996). Several projects have been launched with design team workshops, held at Cranfield or Cambridge universities. Delegates include representatives from the education and property departments, from the school (such as head-teacher and chairperson of the governors). The design team is typically made up of a team leader, architect, quantity surveyor, structural engineer, mechanical and electrical services engineer, interior designer and landscape architect. On several occasions, Essex has used Dr Bill Batty from Cranfield University as the workshop facilitator. Not only has he led the workshop but he has also made presentations to the team about the building design process and about the results of his monitoring of internal conditions in a number of schools, including concentrations of carbon dioxide in class spaces.

The objectives for these exercises have been deliberately limited to the establishment of broad design principles – vehicular and pedestrian access, the general location of the building and playgrounds, form of the accommodation, and broad strategy for the building fabric. The workshops are carefully planned by the facilitator in collaboration with both the design team leader and the project architect.

One such workshop was held at Clare College, Cambridge, in 1997 to design Barnes Farm Primary School, to a programme similar to that shown in Box 12.1. One of the most successful aspects of the event was that the project architect not only held back from bringing to the table a preconceived scheme, but had prepared three 'modelling kits' – comprising a site plan and coloured pieces of card representing the various areas of accommodation required (class spaces, administrative offices, assembly hall and ancillary accommodation). The three breakout groups used these to consider the disposition of the accommodation on the site and the effect on vehicular access, orientation and so on. After about an hour the team reformed and each group presented its ideas to the others. Presentations were followed by a round table discussion about the proposals, chaired and recorded by the facilitator. In consequence, two clear options for the building layout were identified, and the team agreed they would spend the second day examining these options in more depth. These simple modelling kits were universally liked, and the event demonstrated that such an approach need not threaten the traditional role of the architect in producing sketch plans.

The three groups produced quite different solutions – a cruciform plan, a courtyard plan and a radial curved plan. This last was the one selected for further development. A section drawn during the workshop is shown in Figure 12.1. Subsequent to the workshop it was further developed: elevations are shown in Figure 12.2 and the roof plan in Figure 12.3.

The Council's experience of using these design team exercises at the early stages of projects is generally positive, although not universally so. On occasion there has been under-performance, which was diagnosed as being the result of over-familiarity with traditional roles and deep-rooted professional prejudices.

At the end of the Barnes Farm Primary School workshop, the team members were asked whether they had seen any advantages from the exercise and, if so, what these were. The replies were all positive. For example, the landscape architect said 'Having client and ultimate users present helps you to test your preconceptions in a different way'. The team leader said she'd found it very rewarding. The architect said 'It is nice to share problems. Basically it is brainstorming – having everyone around the table is very useful'.

The representative from the education department was equally positive: 'I've heard colleagues argue and constructively discuss positively held views', and went on to express pleasure that there had been 'no riding of hobby horses, no preconceptions'. He also said 'it was useful having different professions challenging different positions'. The county's managing architect said he was pleased that 'people have not just stuck to their own discipline, not just

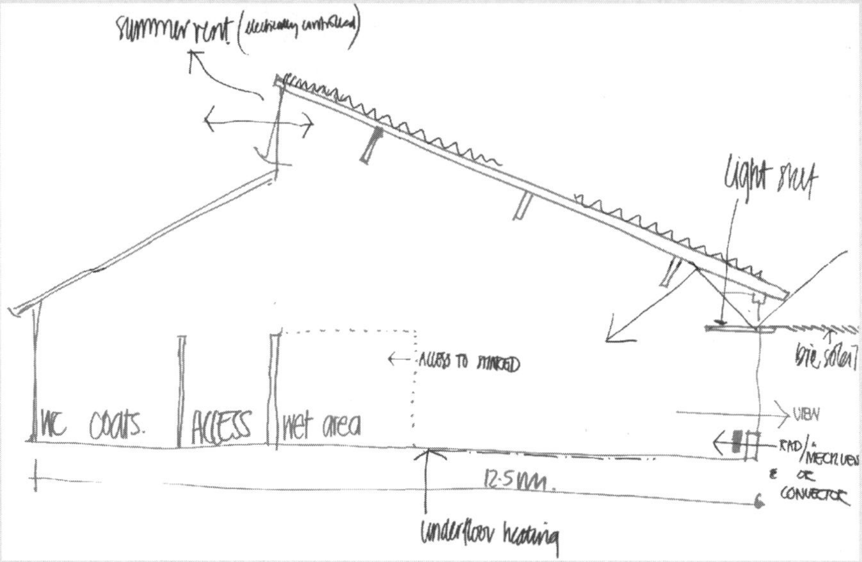


*Figure 12.1 Barnes Farm School: sketch section of the preferred scheme prepared during the workshop.*

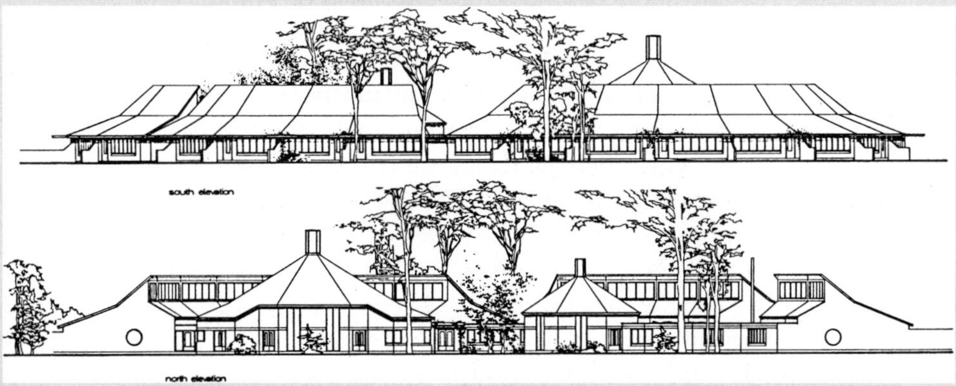


*Figure 12.2 Barnes Farm School: elevations drawn up after the workshop.*

defended their own corner. There has been a spirit of cooperation. No one's bitten back'. He said he hoped the openness demonstrated during the exercise could continue down through the later stages. And he added for the benefit of the client and eventual occupant, 'we've removed some of the mystique seeing the design process in action'.

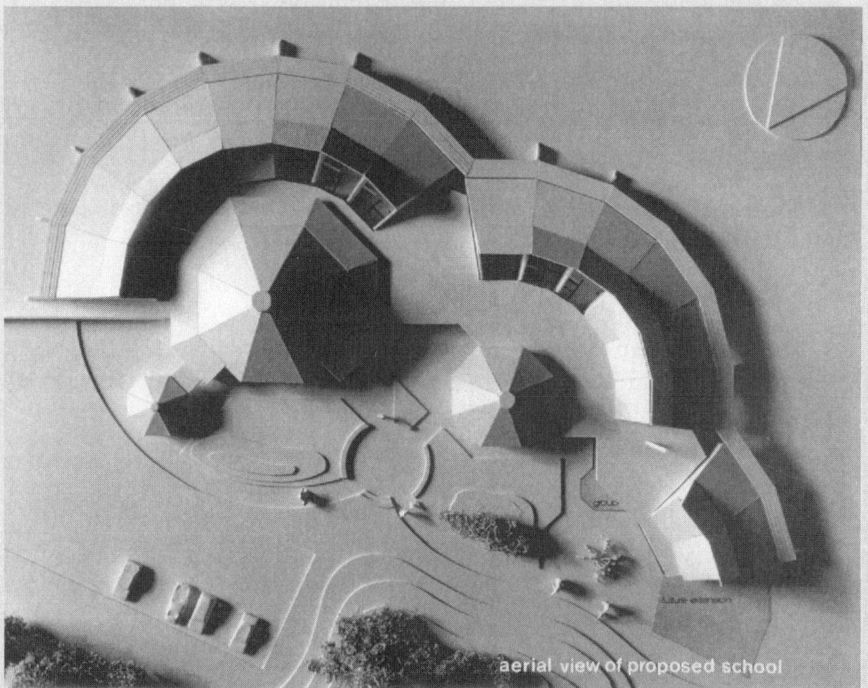


*Figure 12.3 Barnes Farm School: roof plan.*

The future head-teacher was also very positive and said 'I feel happier than if we'd come here to find models of the building already prepared'.

Drawing the event to a close the county's managing architect said they were now mandating the team to go away and work up something. And he noted that one of the benefits of the exercise was the team would be able to justify why it had come to its particular conclusions.

## Key points about design team exercises

The most successful design team exercises are likely to occur where:

- team members are carefully chosen for their willingness to work in this interdisciplinary way
- team members are already familiar with one another – perhaps through working together in a similar workshop or on a project
- team members have experience and expertise in the building type
- the objectives of the workshop are carefully set, the workshop itself is carefully planned beforehand, and the event is professionally facilitated

- delegates include client representatives who can clearly articulate not only what they have had in the past but also what they would like to have in future
- the design exercise is carefully planned beforehand by key team members – such as the facilitator, the team leader and the project architect
- the event itself is run flexibly (rather than rigidly), but also managed so that it achieves its objectives.

The facilitator needs to:

- draw out an agenda rather than imposing it
- achieve a good balance between passive listening and active intervention
- help the team set, and then adhere to, a timetable
- be able to summarize progress from time to time
- assist the team to focus on the objectives of the exercise
- ensure that all participants are given the opportunity to contribute.

The signs of a successfully managed design team exercise are typically that:

- they are conducted in a positive, even a jovial, spirit, punctuated by good humour
- participants interrupt each other constructively, and are prepared to be interrupted
- individuals contribute not only from the perspective of their own discipline but from their wider experience as both designers and users of buildings – in this way the team benefits from sharing the vast pool of expertise across its membership

## Acknowledgements

This chapter is based on work undertaken as part of a research project called 'Achieving quality through interdisciplinary teamwork in design', funded by the Engineering and Physical Sciences Research Council and the Department of Environment, Transport and the Regions under the LINK 'Integration in Design and Construction' research programme.

## References

Austin, S., Steele, J., Macmillan, S., Kirby, P. and Spence, R. (1999) Using training workshops to map interdisciplinary team working. *Proceedings of the CIBSE National Conference 1999*, pp. 38–49.

Batty, W. J., Milford, I. J., Page, B. and Powell, G. (1996) Interdisciplinary working during the strategic phase of building design. In *Proceedings of the CIBSE/ASHRAE Joint National Conference 1996*, Vol. 1, pp. 112–119.

**CHAPTER THIRTEEN**

# Managing an interdisciplinary design team effectively

*Sebastian Macmillan*

## Introduction

This chapter gives advice on how to assemble and manage an interdisciplinary design team in ways that will help ensure that the experience and expertise of its individual members is fully explored and used. It provides definitions of teamwork and describes 'good practice' in forming a design team, running team meetings, managing team dynamics, encouraging open communication, dealing with conflict, and improving team effectiveness. The recommendations originate from a research project undertaken at Cambridge University's Martin Centre for Architectural and Urban Studies. This involved a review of the management literature on project teamwork, and observations of three building design teams during project team meetings. The research team also ran design exercises under controlled conditions (reported in Busseri and Palmer, 2000). The chapter is offered therefore as a research-based digest of good practice, intended to provide practitioners with advice for managing teams, educators with an account of the teamworking skills that are needed in the industry, and researchers with propositions for further testing in the field. A list of references for further reading is appended.

## What teamwork is

Teams are groups of people with complementary skills who are committed to a common purpose and hold themselves mutually accountable for its achievement (Katzenbach and Smith, 1993). Effective teamwork results from:

- well-selected team members,
- coordinated effort and sharing of tasks evenly across the team,

- good communication and open exchange of information, and
- a sense of common ownership of the task at hand and shared responsibility for its achievement.

Teamwork indifference – failing to take action to promote good teamwork – is a strategy likely to result in mediocre performance. Simply bringing together a group of professional designers does not necessarily ensure that they will function effectively as a team or make appropriate decisions (Cooley, 1994). Effective teamwork does not occur automatically and may be undermined by a variety of problems, such as disorganization, poor communication and misunderstanding, and inadequate participation in and procedures for problem solving. Team functioning can be weakened by problems from within the team itself faced by individual members, as well as difficulties linked to the task.

Successful teams are characterized by a team spirit based around trust, helpfulness and – at best – friendliness. And the benefits both to the members of the team members and to the outputs are considered to include (after Denton, 1997):

- a better range of ideas than individuals working in isolation
- improvements in participants' confidence, attitudes and motivation
- greater clarity in expressing ideas through group discussion
- greater optimism – by focusing on positive outcomes and putting less weight on problems
- more effective responses to change – as broader experience helps a team to adapt
- better understanding of the needs of other disciplines and the place of one's own contribution
- better decisions and a more refined solution
- increased personal satisfaction
- more efficient use of resources, especially time.

However, there may also be disadvantages. Team members can waste time and energy in disputes (Hackman, 1987) and some members may opt out of the process, leaving others to do all the work (Erez and Somech, 1996). Varying speeds at which disciplines work may slow the whole group's progress (Batty *et al.*, 1996).

Team members in construction projects are typically drawn from different disciplines. Conventionally this leads to multi-disciplinary teams, comprising representatives from different disciplines. As design issues arise that demand a particular expertise or skill, the person with that skill provides the team with the solution. Interdisciplinary design, by contrast, may be thought of as occurring when problems are solved by the team as a whole, and where members are willing – and indeed are encouraged – to contribute ideas and suggestions in areas that lie beyond their own profession's traditional boundaries.

## Selecting team members

In forming and managing a design team, it is important to consider not only the technical skills required for the task, but also individuals' ability to coordinate their actions and the information they have, and their inter-personal qualities. While it is clearly essential that each team member is competent and experienced in his or her own area of expertise, individual attitudes are also important.

Those who seek to maintain their position by protecting their experience and keeping knowledge to themselves are unlikely to work well as team members, as are those with deep-rooted professional prejudices. Typically in interdisciplinary teams there is greater blurring of the contributions of the different disciplines and it may even be difficult to identify at the end exactly which individuals contributed to which decisions (see Chapter 6). To a large extent, the final outcome is the result of the team process rather than being identifiable with particular individuals.

Teamwork also demands patience and a willingness to explain progress to other team members. Those who prefer to work alone and unaided, and are unwilling to discuss their assumptions, negotiate options and explain solutions, may not function well in teams.

## Leadership and team objectives

Leadership is a critical ingredient for effective teamwork. Effective leaders aim to:

- keep participants focused on the task – making the project as a whole meaningful to individual team members
- offer support to participants where it may be required
- ensure that the team has the resources and information necessary to complete its task
- avoid blaming individuals for problems in the project or in the team
- deal with team members from different organizations who may have varied levels of commitment to teamwork principles
- be aware of participants' loyalties to people (or organizations) in professional groups outside the team – which can complicate team dynamics
- create opportunities for all members to contribute to the task.

One of the key responsibilities of a good team leader is setting team objectives (Adair, 1986). Effective design work in teams becomes more difficult when there are no clear objectives that all members understand. Objectives should be specific and measurable, time-bounded and realistic. Yet they must also be challenging for the team. Giving each member the opportunity to comment on and suggest objectives may encourage a sense of ownership. Clear, shared aims can also help to create a feeling of common purpose, and therefore promote team identity.

## Fostering team identity

Promoting a feeling of identity within the team raises individuals' commitment to their task, which can improve effectiveness. Discussing the team as a unit, as well as its purpose and future development, may help to instil this feeling of identity. Likewise, participants can try to put the good of the team before their own interests or that of their own discipline. Stability and consistency of team membership over the whole project can also help to support team identity and cohesiveness.

Different disciplines may need to be helped to work together to harness the team's collective energy. The team may also have to be helped to identify itself as a team rather than a collection of individuals. Many factors can affect a group's ability to attain its potential team identity, as shown in Box 13.1.

A team's identity and cohesiveness can be challenged if core team constituents change. Often such changes are at the discretion of the collaborating companies, not of the team itself, and arise from commercial or project pressures. Understandable though such changes are, sacrificing an established team can generate additional downstream costs: bringing new team members 'up to speed', plugging gaps in expertise that can open up when team membership changes, and dealing with any antagonism that results from the changes.

## Helping different disciplines work together

Bringing different disciplines together can be difficult, even painful – especially if team members are unsure what to expect from each other, or about each other's capabilities. Moreover, coordination of work and the search for cross-discipline solutions may be limited by the time available for individuals to complete their own work.

**Box 13.1 Ten factors affecting team identity (after Syer and Connolly, 1996)**

1. Openness to discovery and other members' ideas.
2. Getting in tune – shifting the focus from personal needs to the needs of the whole group, and recognizing affiliations both within and outside the team.
3. A clear vision – agreeing where the team wants to be (as distinct from having a purpose or mission statement) and understanding why this is desirable.
4. Establishment of explicit norms indicating what the team is and what it may become.
5. The actions and role of the designated leader and facilitator and of the other members of the team – do they accept joint responsibility and hold each other accountable?
6. The qualities and skills of the team members – and recognizing that these can change and grow.
7. A membership and size of team that matches the task – with sub-group meetings that include some members as and when this is appropriate.
8. Boundaries within the team and those separating the team from outsiders – these determine how contact is made with the team.
9. Resistance on the part of team members to evaluation and self-examination.
10. The willingness of the whole team to take ownership of a problem – rather than framing it as a task for the individual who created or identified the problem.

One goal of collaboration is to ensure that members challenge each other to strive for a common achievement. However, team members need to recognize the human elements involved in cross-culture relationships early on. This can involve identifying social and cultural factors, as well as the technical knowledge and capabilities of those involved.

Teams are at their most productive when they share values, develop a team spirit, absorb the information they need to perform their task, and learn to work well with each other. However, rather than occurring spontaneously, this development process can often be a lengthy one. Design teams may evolve in a variety of ways – some of which involve key issues and identifiable stages characterized by a central issue (see Box 13.2).

Greater trust and familiarity between team members can allow more forceful group behaviour, more willingness to question and challenge in the search for better solutions, and ultimately better designs.

**Box 13.2 Four stages of group development (after Tuckman, 1965)**

1. Forming – where ideas and individuals are tested, relationships are defined and the team agrees on its orientation.
2. Storming – where there is an emotional response to the demands of the task, which may result in hostility and conflict.
3. Norming – where the group becomes more cohesive, and individuals feel able to express their opinions more openly.
4. Performing – where team members adopt the roles they are best suited to, and the team arrives at a solution.

## Team dynamics

In order to achieve their purpose, teams must find ways of working together that allow different, complementary, perspectives to be combined. In successful projects, team members accept each other's perspectives to construct the team's total expertise.

A vital part of how a team behaves is a result of the collective behaviour of its members. It may be difficult to understand how a team works in detail, but it can be helpful to think of it as an amalgam of the personalities, abilities, pressures, loyalties, experience and motivations of all the team's members. People and groups acting outside the team also play a part in shaping how the team and its members behave. Teams are likely to function better if they take time, early on, to talk about the kinds of behaviours that will be valued and how their work will be managed. Individuals' team behaviour can be understood in a variety of terms, including the factors in Box 13.3.

There is a link between how designers view their teamworking and how they view the designs that they produce; the better designers feel about their teamworking, the more positive and pleased they are with their designs. Trust and mutual respect are both vital here. Team members need to know that when they pass the ball their team mates will not drop it. And where people have a different perspective on problems and different priorities, it is necessary to respect what they say and acknowledge that their opinion is valuable. Putting the team interests before their own interests comes more easily to some than to others. Attempts to seek self-gratification from projects – for example the wish to stamp an individual's name on an aspect of the design, rather than accepting that the design reflects the interaction of the team members – can create disputes and tension.

**Box 13.3 Factors affecting individual behaviour in teams (after Belbin, 1993; Hackman, 1987; Driskell et *al.*, 1987; Goodman, 1986)**

1. Team members' personalities.
2. Their mental abilities, skills and expertise.
3. Their current values and motivations.
4. The constraints placed on them by their task or working environment.
5. Their motivations (including rewards) and commitment.
6. Their knowledge and experience.
7. The resources available to them.
8. The opportunities to learn more about (their own and others') roles.

Far from occurring automatically, effective teamwork faces many challenges. In practice, stumbling blocks may be encountered by individual team members, between designers, or between the team and external parties. Alternatively, they may stem from the task itself. Typical problems in team dynamics include:

- no common vision for the project shared by all designers
- making assumptions about what other design team members know
- disagreeing about the extent to which all disciplines take part in design decisions
- failing to identify a stable group of core designers
- failing to coordinate the sharing of information
- the extent to which the client and user groups should be involved in design decisions
- failing to understand and respect how one discipline's work affects the other disciplines
- designers not doing what they say they will, by the time they say they will.

## Communication and conflict

By sharing information between members, a team is able to make best use of its pooled knowledge. Information belonging to people outside the team may be used as an additional resource for teams to draw on. Both forms of information exchange depend on the team's ability to communicate. Yet teams that bring together people from different professional backgrounds, and who use different terminology open up real risks of poor communication and misunderstanding. Such teams need a structure to facilitate joint communication. Groups of strangers are less likely to communicate disagreement (Gruenfeld *et al.*, 1996), while team members are most open to information from those they feel are their equals (Abrams *et al.*, 1990). Team members may also need to make their personal preferences and assumptions explicit (Rentsch *et al.*, 1994) as they strive to understand and trust one another.

Good communication in meetings depends on people's ability to hear what is being said. A meeting room with good lighting and acoustics and adequate space for all participants will help to enhance this. Team members need to speak clearly and strive for clarity, be patient, explain and expand where things are unclear, and listen to others' responses. Active listening can help team members absorb more of what their colleagues from other fields are trying to say. One way to improve listening is for designers to ask questions – focused on what, where, how and why issues – and propose their own ideas when relevant. Being flexible enough to take on board the suggestions of other participants is also important.

When presenting their own ideas, effective listeners aim to use a non-defensive tone of voice, plus body language and eye contact, to show that they welcome feedback. Similarly, formal presentations in meetings may solicit or deter feedback. To encourage feedback, a presentation needs to be well-structured and to include pauses for comments and contributions. Body language and good eye contact can also be used by a speaker to indicate that feedback is welcome.

Conflict can arise when team members have different priorities, goals and/or constraints (Sonnenwald, 1996). The conflict between individual and team goals can present a serious threat to the success of a team (Larson and La Fasto, 1989). Power struggles and unresolved differences in opinion can halt team progress (Bolmar and Deal, 1992). Consequently, identifying and resolving conflict is an inevitable part of effective team process. Teams may not develop unless individual members take risks, some of which involve conflict. Conflict can be a source of energy and creativity and, at best, an opportunity to discuss individual differences and concerns and integrate them into common goals.

Contentious topics are sometimes deliberately skated over in meetings. On occasion, they are resolved using authority relationships – with one team member exercising authority to insist their preferred action is followed. In either case, the root causes of conflict are not faced or resolved and the opportunity to address them head on is lost. The key lies in resolving, not avoiding or dealing superficially with, conflict when it arises.

## Reflection and self-assessment

One of the first steps towards better teamwork is for team members to discuss and agree what exactly 'teamwork' means to each of them. Defining teamwork can help to shape how team members work together and what they strive towards, setting common expectations of behaviour for all members. Such a discussion might range across at least the following issues:

- the importance of identifying individual roles and responsibilities and defining goals and concepts
- how to get the best from design team meetings and encourage participation
- the importance of effective communication between team members
- how teams develop, and how to harness the collective energy and expertise of the members
- how to foster team identity, improve team effectiveness and cope with conflict.

Design teams rarely recognize the value of talking about how they work together – the 'process' of teamwork. They tend to focus only on the product they are working on. Sometimes the reluctance to discuss the process even extends to planning how they will approach a project. Yet team development can be enhanced by self-reflection. Reviewing from time to time the interaction between team members, the clarity of communication, and the effectiveness of collaboration, can lead to identification of problem areas and potential improvements. Such self-reflection can address different priorities among team members and the factors that cause a team to stray from its objectives.

One way to do this is for a team to assess from time to time whether it is genuinely working as a single team, focusing on team objectives, collaborating and communicating effectively, and taking account of everyone's suggestions. At best, this kind of evaluation of the team's past performance will identify deficiencies and address how best to improve future performance.

Criteria for evaluating the team's performance might include whether (after Ancona, 1987 and Neumann *et al.*, 1995):

- the team's output meets or exceeds the standards of the client, the users or the public
- milestones and deadlines are met
- members are more satisfied than frustrated by the group experience
- members learn to work successfully together over time
- the team achieves its aims in the most efficient way

- individuals focus on what is good for the team more than what is good for them personally

Team effectiveness should also be discussed in terms of the quality of work produced. Quality can refer to the features of a product that affect user satisfaction, or doing the work in the right sequence and on time, or even the viability and clarity of documentation and plans. Teams may benefit from discussing and agreeing on a shared meaning and level of quality.

## Design team meetings

Meetings are an essential component of effective teamwork and are probably the most important forum where designers work together. They serve two main functions:

- a social purpose – in which team members become familiar with one another and one another's ways of working, and
- a business purpose for conveying information, agreeing actions and taking decisions.

Meetings can be extremely valuable as a deadline for individuals to produce information needed by other team members. They can, however, sometimes be a waste of time. Unless there are overriding social reasons for holding a meeting, they should be held only where they will result in decisions and actions. Simple conveying of information between team members is best undertaken through other routes – email, telephone or by post.

Meetings can serve two general purposes:

- sharing information among team members, for example about progress against the programme – leading to the team agreeing actions
- decision making – including explorations of alternatives, assessments of options and finally selection and choice.

Design team meetings sometimes combine both purposes – informing and decision making – and it is important to avoid ambiguity between them. The structure and preparation for the meeting should clarify what is on the one hand information distribution and open-ended informal discussion, and on the other hand the issues that require firm decision making.

Forward planning of meetings according to a schedule is usually preferable to an *ad hoc* timetable. Pre-planning helps to ensure that participants keep the dates free of other commitments, and it also helps participants to keep track of deadlines and milestones. However, overzealous scheduling of meetings too far in advance can result in a need to change dates to suit unforeseen circumstances, and may be counter-productive.

Meetings in the construction industry are often run informally, lacking a formal agenda, circulated papers and drawings, and – at worst – are poorly chaired. An agenda circulated among participants in advance is essential for every meeting. Prepared in consultation with all team members, who are given the opportunity to add agenda items, it sets out the issues to be addressed so that participants know what to expect and can prepare the information they need. An agenda defines the issues to be addressed at the meeting and the problems that need to be resolved. Without an agenda, key decisions may be overlooked, and meetings can end up focusing on relatively minor technical issues that may be of immediate concern but fail to address more strategic issues.

Pre-circulating drawings and reports is also essential. It is all too common in design team meetings for vital drawings and/or information to be brought to the meeting on the day rather than circulated in advance. Failure to pre-circulate reports and distribute up-to-date drawings prior to a meeting gives other team members inadequate time to absorb them. Drawings are a particular case. Because they are large and costly to reproduce it can happen that single copies are brought to a meeting and pinned up or spread over a table – with the expectation that team members can assimilate them and make any necessary decisions at the meeting. Sometimes this is the result of time pressures – drawings are hot off the press and there is too little time to circulate them in advance. Occasionally it is deliberate – a way of preventing the team members having a full debate about the fabric or services. But such behaviour soon results in frustration among team members and a loss of trust. Failure by individuals to prepare for a meeting properly in advance prevents the team from getting the full value from meeting time.

Chairing a meeting is a skill which needs to be developed by all those who intend to lead a team. A chairperson at a meeting will ensure that every agenda item is covered and keep the meeting to time, including bringing it to a timely conclusion rather than just letting participants drift away towards the end. Excessively long meetings with insufficient refreshment breaks can lead to fatigue and loss of attention among participants. Effective chairing also involves ensuring that all members have the opportunity to participate, particularly those who are unaccustomed to early involvement in design and who may not push their own views forward unsolicited. Meetings which are well-chaired and highly productive raise enthusiasm and satisfaction among team members (Jones, 1992).

Jumping straight into the agenda as soon as members arrive, and closing without any kind of review, can also act against group identity. Warming up in meetings is therefore important – as a way to remind team members what the team is about, and what you are aiming to achieve. Typically this could involve an opportunity for participants to introduce themselves, a restatement of the 'mission statement', project objectives, and deadlines for the project, and a summary of progress so far. This will serve as a reminder to re-acclimatize the group to working as team; that each person is necessary to complete the task; that they are dependent on (and accountable to) the rest of the group; and, for those who are simultaneously engaged on other projects, to focus afresh on the project in hand.

All participants at meetings should be encouraged to keep their own notes. But, in addition, formal minutes should be taken either by the team leader or by some other person present. At the very least, the formal minutes should record for each agenda item the decisions taken, the actions agreed and who is to undertake them. Ideally, the debate leading to individual decisions should also be recorded as this will provide explanation and justification for the decisions – should this be required in the future; for example for audit purposes. Minutes need to be distributed as soon after the meeting as possible. Minutes of past meetings should be available to the team for reference.

A clear disadvantage of all team members attending meetings for the whole duration of a project is that it absorbs design resources. It can also happen that some design issues – such as specific technical details – are relevant to only some of the team members, leaving other members unable to contribute. Meetings called for just some team members can resolve this, where the participants are chosen according to the objectives of the meeting. The benefits from sub-teams chosen in this way need to be offset against the advantages of exposing all team members to the concerns of other disciplines. Being present during specialist discussions may help to build respect for other designers. Witnessing decision making for other aspects of the design may also cut the likelihood of making changes that call for complex re-working – so *saving* time in the long run. If sub-team meetings are used, the outcomes need to be transmitted to all the other members of the team.

## Summary of key issues

- Understanding team dynamics needs consideration of how teams work together and develop over time, how to make the most of the team's collective energy, and how to foster team identity.

- In order to achieve their purpose, teams must find ways of working that allow different, complementary perspectives to be successfully combined.
- The dynamics of a team are ultimately dependent upon the behaviour of individuals – what they do and how they relate to other people. But teams seldom recognize the value of discussing the *process* of teamwork – they typically focus almost exclusively on the *product* they are working on.
- Combining insights from a variety of disciplines is complicated and can be painful. Recognizing human elements involved when professionals from different cultures come together can help relieve some of this pain.
- There is an important link between how well team members work together, and the designs they produce.
- The process of team development can take a significant amount of time, during which a team may evolve in a variety of ways. Every team (even teams with very similar tasks) needs to develop its own strategies to make effective teamworking possible.
- A team's effectiveness and satisfaction may rise if its identity is strengthened – by ensuring that all members understand and buy into its purpose and approach to work.
- When a team fuses successfully, its designs are not identifiable as the property of a single individual. Instead, they reflect the interaction of, and contributions from, all members.
- Teams should review and assess their own processes periodically in terms of participation, clarity of communication, and the opportunities to promote team identity.
- Effective teamwork does not occur automatically and may be challenged or weakened by problems that come from the team itself, problems facing by individual members, or difficulties that result from the task.

## Acknowledgements

This chapter draws extensively on an unpublished booklet called *Building teams: working together in building design* by Jason Palmer, Michael Busseri and Sebastian Macmillan. It was the key output from a research project called 'Achieving quality through interdisciplinary teamwork in design' funded by the Engineering and Physical Sciences Research Council and the Department of Environment, Transport and the Regions under the LINK 'Integration in Design and Construction' research programme.

## References and further reading

Abrams, D., Wettherell, M., Cochrane, S., Hogg, M., and Turner, J. C. (1990) Knowing what to think by knowing who you are: self-categorization and the nature of norm formation, conformity and group polarization. *British Journal of Social Psychology*, **29**, 97–119.

Adair, J. (1986) *Effective Teambuilding: How to Manage a Winning Team*. Pan, London.

Alderfer, C. P. (1987) *An Intergroup Perspective on Group Dynamics*. Prentice Hall, Englewood Cliffs, NJ.

Ancona, D. G. (1987) Groups in organisations. *Review of Personality and Social Psychology*, **9**, 207–230.

Arditi, D. and Gunaydin, H. M. (1995) Total quality management in the construction process. *International Journal of Project Management*, **15**, 235–243.

Arup, O. (1995) Key speech, *The Arup Journal*, **20**, 34–37.

Baden Hellard, R. (1993) *Total Quality in Construction Projects*, Thomas Telford, London.

Batty, W. J., Milford, I. J., Page, B. and Powell, G. (1996) Interdisciplinary working during the strategic design phase of building design, *Proceedings of the 1996 CIBSE/ASHRAE Joint National Conference*, Harrogate, 112–119.

Belbin, M. (1993) *Team Roles at Work*. Butterworth-Heinemann, London.

Bettenhausen, K. L. (1991) Twenty-five years of groups research: what we have learned and what needs to be addressed. *Journal of Management*, **17**.

Bolmar, L. G. and Deal, T. E. (1992) What makes a team work? *Organisational Dynamics*, **21**, 34–44.

Busseri, M. and Palmer, J. (2000) Improving teamwork: the effect of self assessment on construction design teams. *Design Studies*, **21**, 223–238.

Cooley, E. (1994) Training an interdisciplinary team in communication and decision making skills. *Small Group Research*, **25**, 5–25.

Cross, N. and Clayburn Cross, A. (1995) Observations of teamwork and social processes in design. *Design Studies*, **16**, 143–170.

Davidson, R. (1993) An instrument for measuring meeting success. *Information and Management*, **32**, 163–176.

Denton, H. G. (1997) Multidisciplinary team-based project work: planning factors. *Design Studies*, **18**, 155–170.

Driskell, J. E., Hogan, R. and Salas, E. (1987) Personality and group performance. *Review of Personality and Social Psychology*, **9**, 91–112.

Durcan, J. and Oates, D. (1994) *The Manager as Coach: Developing Your Team for Maximum Performance*. Financial Times/Pitman Publishing, London.

Erez, M. and Somech, A. (1996) Is group productivity loss the rule or the exception? Effects of culture and group-based motivation. *Academy of Management Journal*, **39**, 1513–1537.

Fleming, Q. W. and Koppelman, J. M. (1996) Integrated project teams: another fad... or a permanent change? *International Journal of Project Management*, **14**, 163–168.

Gersick, C. J. G. (1988) Time and transition in work teams: toward a new model of group development. *Academy of Management Journal*, **31**, 9–41.

Goodman, P. S. (1986) *Designing Effective Work Groups*. Jossey-Bass, London.

Grossman, S. (1997) Turning technical groups into high performance teams. *Research-Technology Management*, **40**, 9–11.

Gruenfeld, D. H., Mannix, E. A., Williams, K. Y. and Neale, M. A. (1996) Group composition and decision-making: how member familiarity and information distribution affect process and performance. *Organizational Behavior and Human Decision Processes*, **67**, 1–15.

Hackman, J. R. (1987) *The Design of Work Teams*. Prentice Hall, Englewood Cliffs, NJ.

Hatcher, L. and Ross, T. L. (1991) From individual incentives to an organization-wide gainsharing plan: effects on teamwork and product quality. *Journal of Organizational Behavior*, **12**.

Jones, S. (1992) *The Human Factor: Maximising Team Efficiency Through Collaborative Leadership*. Kogan Page, London.

Jones, M. C. and Harrison, A. W. (1996) IS project team performance: an empirical assessment. *Information and Management*, **31**, 57–65.

Katz (1982) cited in Ancona, D. G. (1987) Groups in organisations, *Review of Personality and Social Psychology*, **9**, 207–230.

Katzenbach, J. R. and Smith, D. K. (1993) *The Wisdom of Teams*. Harvard Business School Press, Boston, MA.

Keck, S. L. (1997). Top management team structure: differential effects by environmental context. *Organization Science*, **8**, 143–156.

Larson, C. E. and LaFasto, F. M. (1989) *Teamwork: What Must Go Right; What Can Go Wrong*. Sage, London.

Latham, M. (1994) *Constructing the Team. Final Report of the Government/ Industry Review of Procurement and Contractual Arrangements in the UK Construction Industry*. HMSO, July 1994.

Lawson, B. (1983) *How Designers Think*. Butterworth Architecture, London.

McGrath, M. (1991) *Multi-Disciplinary Teamwork*. Avebury, Aldershot.

Morrow, P. C. (1997) The measurement of TQM principles and work-related outcomes, *Journal of Organizational Behavior*, **18**, 363–376.

Muir, T. and Rance, B. (1995) *Collaborative Practice in the Built Environment*. E. & F. N. Spon, London.

Mullen, B. and Baumeister, R. F. (1987) Group effects on self-attention and performance. *Review of Personality and Social Psychology*, **9**, 189–206.

Neumann, J. E., Holti, R. and Standing, H. (1995) *Change Everything at Once!* Management Books, Oxfordshire.

Newcombe, R. (1996) Empowering the construction project team. *International Journal of Project Management*, **14**, 75–80.

Rentsch, J. R., Heffner, T. S. and Duffy, L. T. (1994) What you know is what you get from experience: team experience related to teamwork schema. *Group and Organisation Management*, **19**, 450–474.

Sonnenwald, D. H. (1996) Communication roles that support collaboration during the design process. *Design Studies*, **17**, 277–301.

Stokes, J. (1994) Problems in multidisciplinary teams: the unconscious at work. *Journal of Social Work Practice*, **8**, 161–167.

Sutton, R. I. and Hargadon, A. (1996) Brainstorming groups in context: effectiveness in a product design firm. *Administrative Science Quarterly*, **41**, 685–718.

Syer, J. and Connolly, C. (1996) *How Teamwork Works: the Dynamics of Effective Team Development*. McGraw-Hill, London.

Tajfel, H. (1982) *Social Identity and Intergroup Relations*. Cambridge University Press, Cambridge.

Tjosvold, D. (1984) Cooperation theory and organizations. *Human Relations*, **37**, 743–767.

Tuckman, B. W. (1965) Developmental sequence in small groups. *Psychological Bulletin*, **63**, 384–399.

Walsh, J. P., Henderson, C. M. and Deighton, J. (1988) Negotiated belief structures and decision performance: an empirical investigation. *Organizational Behavior and Human Decision Processes*, **42**, 194–216.

Whitfield, J. M., Anthony, W. P. and Kacmar, K. M. (1995) Evaluation of team-based management: a case study. *Journal of Organisational Change Management*, **8**, 17–28.

## CHAPTER FOURTEEN

# Towards an interdisciplinary design process

*John Steele and Sebastian Macmillan*

## Introduction

In this chapter we report on the results of a research project undertaken at Cambridge University's Martin Centre concerned with mapping the early stages of the design process. These stages are a time when decisions with fundamental and extensive effects on appearance, performance and costs are made, and when the design team can add the greatest value to a project. It is often a vibrant, dynamic and creative period, but it can also be chaotic as a new team comes together for the first time, the client's requirements need to be refined, and there are many uncertainties about project costs and feasibility. This research focused on the development of a map or programme to assist the interdisciplinary design process during the early stages as a means to obviate the need for the design team to reinvent the process each time, and to help ensure the design outcome fully reflects the combined expertise of the project participants.

The early stages of the design process are not well understood (Hedges *et al.*, 1993) and, for designers in practice, available advice is mostly limited to lists of deliverables; there is little or no guidance on what should be done and how it should be achieved (Macmillan *et al.*, 1999). As a result, design teams often have very little structure or common focus, which can be a source of confrontation between individuals. If teams are to be effective, individual members need to collaborate by contributing ideas and misgivings, by listening as well as speaking, by negotiating as well as demanding, and by exchanging information and perceptions. Collaboration relies upon team members supporting one another in both addressing questions and picking-up and building upon each other's lines of thought (Valkenburg and Dorst, 1998).

This research starts from the premise that interdisciplinary teamwork in design can be supported and maintained at the concept stage through the introduction and sharing of a flexible framework that outlines the events through which the design activity is likely to pass. Like other plans of work and process protocols, such a framework is intended to assist the team by promoting collaborative design development and acting as a basis for coordination and negotiation among team members (Peng, 1999). There is some evidence to suggest that designers in possession of such a framework are better able to focus on the demands of a problem than those without (Archer, 1984).

Our literature search identified many past empirical investigations into the activity of design, some carried out in the work environment (for example Hales, 1987; Bucciarelli, 1988; Minneman, 1991) and others in experimental settings (for example Radcliffe and Lee, 1990; Tang and Leifer, 1991; Christiaans and Dorst, 1992). Most have focused on individual designers (Cross and Clayburn-Cross, 1996). Yet, in practice, most construction projects of any significance involve collaboration between different disciplines working as a team. There is far less design research which has addressed the added complications that arise from interdisciplinary team work.

The literature review identified a range of issues about how teams of designers function at the conceptual stage of the design process. These issues were investigated through direct observation of practising designers, case study interviews, archive analysis, and experimental workshops involving designers from all of the relevant disciplines. Analysis of the data led to the development of a tentative design framework comprising 12 main activities organized in five phases. This 'model' of design activity was then turned into a prototype interactive Web-based system intended to provide decision support for designers. The prototype was field tested on a live project in industry.

## Case study investigations

Thirty design professionals were interviewed about their design activity during the conceptual design phase of nine design and construction projects. To triangulate the interviewees' descriptions and perceptions, archived documents – meeting minutes, early design drawings, project design notes and concept design reports generated over the course of each project – were analyzed. Table 14.1 provides a summary of the case study details. For each one, the various sources of data were compiled, collated and cross-referenced. A synopsis was prepared of each which

*Table 14.1 Comparison of case study projects.*

| Project | Project type | Client type | Phase duration | Building cost (budget) | No. of individuals interviewed | No. of documents studied | No. of tasks described |
|---|---|---|---|---|---|---|---|
| A | Airport pier | Airport authority | 12 weeks | £21.5m | 3 | 4 | 36 |
| B | Airport terminal (refit/extension) | Airport authority | 12 weeks | £51m | 2 | 3 | 41 |
| C | Airport terminal (refurbishment) | Airport authority | 8 weeks | £12m | 4 | 4 | 30 |
| D | Office building refurbishment | Pharmaceu-tical organisation | 6 weeks | £16.7m | 3 | 5 | 32 |
| E | Laboratories, offices (new build) | Pharmaceu-tical organisation | 20 weeks | £137m | 4 | 8 | 58 |
| F | Laboratories, offices (new build) | Pharmaceu-tical organisation | 12 weeks | £30m | 3 | 6 | 40 |
| G | Corporate offices (new build) | Property developer | 8 weeks | £28m | 5 | 9 | 42 |
| H | Operations centre | Rail company | 5 days | £20m | 3 | 3 | 27 |
| J | Production facility/office headquarters | Private client | 2 weeks | £2.5m | 3 | 2 | 45 |

included a listing of the design tasks that were involved. An example of one case study – Project F – is given in Box 14.1.

## Box 14.1 Project F case study synopsis

The client had produced an initial briefing document which outlined requirements for further development of an existing site, and used this to establish the initial budgetary approval required to proceed with the proposal to concept study stage. The initial briefing document was reasonably detailed as the client was experienced and had a firm idea of what was required. The brief was based on the client's existing facilities and laboratory environments. Initial project meetings involved a number of key individuals from the predesignated project team, the client's user team (the heads of department who would be eventually using the facility), and the client's private quantity surveyor. These meetings allowed the design team to hold initial interviews and discussions with the users of the facility to discuss their requirements and generate a provisional schedule of accommodation. There were several key reasons for holding this session.

- Meet the client representatives (user team).
- Discuss and understand the user requirements.
- Identify and verify any design drivers.
- Identify the preliminary design constraints.
- Gather information to allow production of a provisional schedule of accommodation.
- Undertake a walk around the site and take notes and photographs on the existing facilities.

The client provided details of the number and type of scientists that were to inhabit the facility, their desk space and general furniture requirements, and the number of fume cupboards needed. Information about the number of scientists required to work in each particular area, and the need for interaction between certain types of scientist was provided. The size of a standard module for four people was calculated and used to generate the overall size of floor plate and form of the building. A second meeting was then held to allow the design team to develop their understanding of the form of the building through a deeper understanding of the aspirations, culture and working relationships of the users, and the attributes that were of value to the project stakeholders.

This second meeting involved the design team leader, a principal architect and the project economist, in addition to the client-designated user team (made up of the laboratory unit heads) and senior client representatives. Several key issues were discussed at this meeting, with the main aim being to gather further information on the aspirations of the client and the facility's users. Discussions involved:

- collecting further information on the building's functional requirements
- gaining an idea of where on the site the building should be located. (The client had a reasonably good idea of where the building should be located at this point, so a brief master planning exercise was undertaken to evaluate alternative locations)
- assessing the client's preferences for spatial arrangements and building form
- further understanding the nature of the company culture and need for interaction of employees within the facility
- assessing preference for modular or craft approach
- gaining a general understanding of the user team's aspirations for the building.

Broadly speaking, this session allowed the design team to interpret what it was that they were aiming to produce. The session resulted in several key decisions being taken as to the way forward for the project. In turn, this allowed the design team to produce preliminary (elemental) cost targets based on benchmarks, lists of primary, secondary and tertiary requirements (in terms of value to the client) for the facility, and to generate a project mission statement. Additionally, a value tree representing client requirements was generated. Following this, the design team produced a short questionnaire for completion by each of the client's laboratory unit heads. It was organized around the discussions held at the initial meeting, and was reinforced with visits by the design team, their purpose being to understand and develop the user-designated functional requirements more fully. The questionnaire replies were followed up by discussions with the client representatives. Through these discussions it became apparent that, owing to the need to accommodate future changes in the facility, a modular system was needed.

A specific list of functional requirements was produced and developed into a document outlining the preliminary design strategy for client approval. The form of the building evolved in response to comments from the client user team and the work of the architects. Optimum room shapes were developing around client dimensions, and these were used, in conjunction with the continuously evolving schedule of accommodation, to devise overall spatial arrangements. The services and structural elements of the design philosophy were also being developed in conjunction with the architectural philosophy. Once the numbers of occupants and the location of particular spaces were firmed up, services engineers undertook preliminary sizing of the plant room and its location. They also developed a vertical and horizontal distribution strategy. Structurally, the column arrangement was almost fixed by the modular approach. However, a decision was taken to expose the columns externally and embed them within the depth of the services corridors internally.

Several options for the building form and spatial arrangement were generated by the design team, and the architect began to produce sketch options (based on past precedents) of the general building form, while simultaneously exploring alternative cladding types. Discussions of building facades ensued, involving the architect and the building services engineers, who identified options for cladding systems based on past experience within the design office. The client was shown photographs of existing buildings and visited a number of other facilities to review facade options. The client also asked for a full-height glazed atrium as an entrance to the building.

Meanwhile, the mechanical and electrical services engineers had concentrated on gathering all the information available on the existing site infrastructure in a bid to assess whether existing utility capacities were sufficient to serve the new building, while also making informed guesses about services options. They then began to generate options for the servicing strategy based on the initial client briefing document, basic requirements from 'rules of thumb', and their own previous experience of similar facilities.

The architects devised two separate options for the overall building, while the civil/structural engineers generated options for the superstructure, foundations, site infrastructure and car parking arrangements. Their work had initially involved site surveys and recommendations about services diversions. A design decision was made about the optimum location of the plant room, allowing each of the disciplines to advance their general philosophies in parallel. The mechanical services philosophy was developed in conjunction with the building architecture (form, spatial arrangements, floor layouts and preliminary facade and cladding proposals) and the structural requirements and options were further refined. This allowed the mechanical philosophy to be firmed up, and the electrical requirements to be incorporated into an electrical infrastructure philosophy.

Finally, a graphic designer worked with the architect to prepare sketches of internal and external views, and then rendered perspectives. These, together with all the supporting information generated by the engineers, surveyors, architects, and other specialists, were incorporated into the concept study document for submission to the client. This brought the concept stage of the project to a conclusion.

## Developing block diagrams of design activities

Block diagrams were produced to represent each of the case studies. Figure 14.1 illustrates the diagram constructed from the description of case study F. The block diagrams were then compared and collated to identify common patterns across all cases.

The number of tasks that were identified in each project varied between 27 and 58. These variations were due partly to differences in the design teams' perceptions and interpretations, and partly to project differences. While some activities were project-specific, certain tasks were common across every project studied.

There was little, if any, explicit recognition of iteration within or across the identified tasks. Designers' accounts typically implied systematic and linear progression in the design activity. Many interviewees did acknowledge that iteration had occurred across tasks, but they could not recall how this iteration had occurred. Every project, however, had involved a period of understanding the client's requirements prior to any generation of concepts. Although the descriptions from interviewees suffered from the weakness of post-rationalization, they provided valuable insights into how designers perceive the concept design phase. Together with the project documentation and other published process maps, they contributed to the development of the generic design framework for concept design.

## A design framework for concept design

The initial design framework for concept design drew on both the block diagrams built up for each case study, and from a review of published process maps of the concept stage from construction and other sectors (Macmillan *et al.*, 1999). A comparison of some published maps is shown in Figure 14.2. A process of iteration led to the identification of terminology, levels and hierarchical structure of a framework for concept design.

Following a trial use of the framework by three design teams during an AMEC training workshop on interdisciplinary design (Steele *et al.*, 1999), it was further revised and refined. It now comprises a hierarchical model with five levels of definition, four of which are generic. At the first level it covers simply 'conceptual design', the second level comprises two stages, the third contains five phases, and these are further subdivided into 12 activities in level 4. At levels below this, we anticipate that it will be necessary to move away from generic definitions and address issues that are specific to a sector – such as schools, offices, hospitals and so on (Figure 14.3). Definitions of the 12 activities in level 4 are given in Table 14.2.

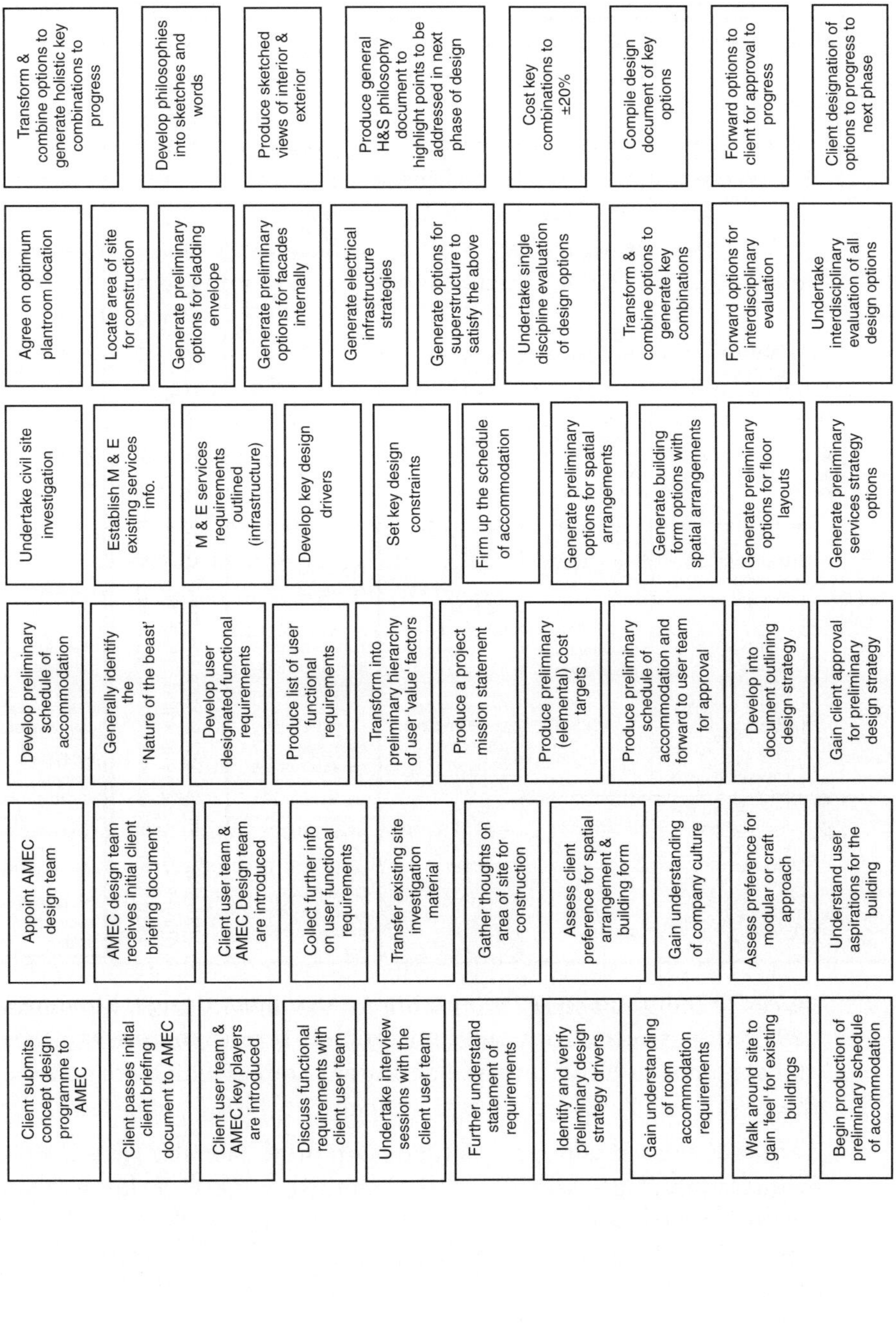


*Figure 14.1 Block diagram of task description derived from Project F.*

| Literature review (section of larger comparison – Macmillan *et al.*,1999) | | | | Workshop descriptions | Case study projects | Proposed generic model |
|---|---|---|---|---|---|---|
| **Hubka (1982)** | **Pahl and Beitz (1988)** | **Cross (1989)** | **Jones (1992)** | **Proposal for terminology** | **Task grouping** | **Actual framework terminology** |
| | | | | Understanding the briefs requirements | The clustering of tasks from the case study projects was undertaken with the aspects of the prior investigations in mind. Thus the genesis of the preliminary design framework from the case study tasks was grounded in the authors synthesis of available data | Specify the need |
| | | Clarifying objectives | | Assessing and developing design requirements | | Assess the requirements |
| | Identify essential problems | | | Identifying design drivers and constraints | | Identify essential problems |
| Establish function structure | Establish function structures | Establishing functions | Design situation explored | | | Develop the requirements |
| Establish technical process | | | | | | |
| Apply technical systems and establish boundaries | | Setting requirements | Problem structure perceived and transformed | Prioritising factors and constraints | | Set key requirements |
| Establish groupings of functions | | Determining character-istics | | | | |
| | | | | | | Determine project characteristics |
| Establish functional structure and represent | | | Boundaries located, sub-solutions described and conflicts identified | | | |
| Establish inputs and modes of action | Search for solution principles | Generating alternatives | | Generating concepts | | Search for solutions |
| Establish classes of function carriers | | | | | | |
| Combine function carriers and examine relationships | Combine solution principles | | | Grouping/com-bining solution concepts | | Transform and combine solutions |
| | Select suitable combinations | | Sub-solutions combined into alternative designs | Selecting suitable options | | Select suitable combinations |
| Establish basic arrangement | Firm into concept variants | | | | | Firm into concept variants |
| | | | Alternative designs evaluated and final design selected | | | |
| | Evaluate against technical and economic criteria | Evaluating alternatives | | | | |
| | | | | Evaluating/ choosing options | | Evaluation and choice of alternatives |
| | | Improving details | | Developing, improving and reviewing of final options | | Improve details and cost options |

*Figure 14.2 Sources of the preliminary framework (at the activity level) for concept design.*

Once complete, the framework and the activity level definitions were further discussed with practising design professionals. They were also evaluated during a second AMEC interdisciplinary design workshop, where it emerged that, although the phases and activities were representative of the conceptual design process, the framework did not account for the social interaction that is required to maintain teamwork.

*Table 14.2 Definitions of level 4 framework activities.*

| Activities | Definition |
|---|---|
| Specifying the business need | The client recognizes that the only way to satisfy the business need is to construct a building. The need may be expressed formally in a design brief or verbally. The design team must appreciate and understand this need before attempting to address it. This can usually be achieved by gathering the information available and then generating a project mission statement to define broadly what is required. |
| Assessing functional requirements | This activity involves taking the specified business need and eliciting and recording the requirements for functionality and performance. If the client's own requirements are not met, the design solution is unlikely to be acceptable. |
| Identifying essential problems | Here the design team should identify the constraints of the problem both to clarify it and to promote some solution-focused thinking. That a new design is required shows that there are problems with the products or systems already available; these problems should be identified and used to guide the design by setting some design drivers and constraints. |
| Developing functional requirements | The design team should attempt to extend the acceptable solution boundary. This can be achieved by identifying the 'real' users of the facility or system and questioning them about their needs and wants. This may stimulate innovative solutions. |
| Setting key requirements | This activity involves the production of a list of all the requirements that have been both stated by the client and introduced by the design team. Each should then be defined in a few words, and in ways that give the design team direction while still being broad enough to allow a wide search space. Similar requirements should be combined, while those that are unrealistic should be discarded or reassessed. The requirements that remain should be set as key to the project. |
| Determining project characteristics | The pre-set requirements define the boundaries of the search space. Not all requirements can be met equally and some will inevitably be more important than others. Working with the client, the team should rank or prioritize the requirements in order of their perceived value, and thus importance, to the success of the project. This will define the project characteristics and will set the value datum against which the conceptual design proposals will be assessed. |
| Search for solution principles | This activity is where initial attempts are made to generate solutions. These solutions can be developed based on requirements or abstract ideas. Designers should be creative and uninhibited in proposing solutions. The key here is to combine creative thinking with experience and prior knowledge. Externalizing ideas opens them up to critical examination, and also ensures that ingenuity is recorded, however useless or unrelated it may initially appear. |

| | |
|---|---|
| Transform/ combine solution principles | At this point a number of concepts should have been generated. These, however, may well be unusable in their present form, and this period of design activity should concentrate on developing, transforming and combining individual proposals in a bid to mould them in to a number of usable proposals, at system or sub-system level. |
| Selecting suitable combinations | The number of solution concepts must be reduced as early as possible. However, care must be taken not to dismiss valuable concept principles before the opportunity to combine them with other concepts to generate an advantageous overall solution is lost. There is no fail-safe procedure for this, but the decision should be democratic. Elimination of totally unsuitable proposals must be undertaken. After this, preference should be given to those remaining solutions that are patently better than the rest. |
| Firming up into concept proposals | Those concepts that remain may satisfy the requirements superficially. However, the generation and selection procedure to date may well have revealed gaps in information about important elements of the design that mean that a reliable evaluation is impossible in their present state. Attention needs to be paid to coordination across discipline boundaries. More detailed information need only be gathered for promising proposals. |
| Evaluation and choice of alternatives | This activity involves the solution proposals or concept variants being evaluated by the interdisciplinary design team so as to provide an objective basis for decisions. The views of individuals must be supported by evidence or explanation that convinces the team of their validity and justification. Evaluation procedures allow the concept variants to be gauged relative to one another, and for the extent to which they meet the client's requirements. |
| Improve details and cost options | This activity requires the improvement of details and the costing of the proposals. The costing of proposals should be an ongoing exercise throughout the design activity, but at this juncture a detailed costing of the proposals is imperative. The chosen proposal should be taken to a level that allows the critical unknowns to be understood sufficiently so they pose little or no risk to the subsequent development and success of the project. The pinch points should be detailed enough to ensure that coordination can be facilitated in the later stages of design. This activity, along with the entire conceptual design phase, is complete when the chosen proposal is documented in a way that the client can fully understand it and, as such, agree that with further development it will satisfy the client's business needs. |

Fortunately, another research project at Cambridge (see Chapter 13) had explored these issues and had generated complementary material, enabling a team maintenance component to be introduced into the map.

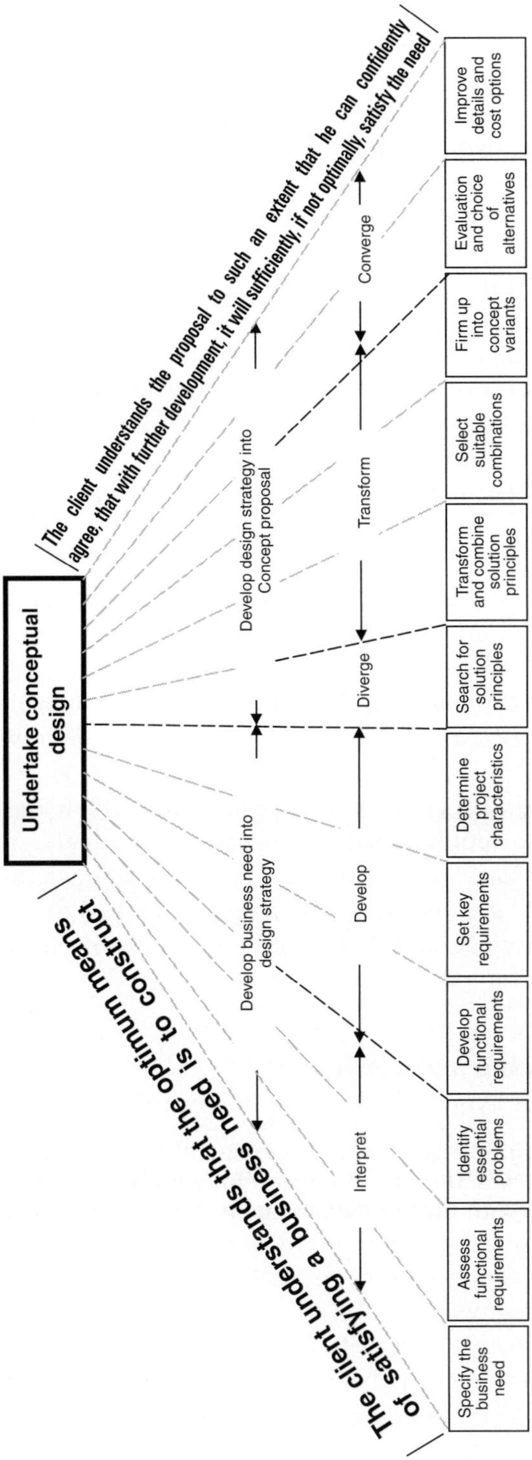


*Figure 14.3 The revised version of the preliminary framework for concept design.*

## Delivery of the framework as a Web-based design support system

Design is a richly iterative activity and one of the disadvantages of a paper-based model of the design process is the sequential nature implied by paper. This itself may be a disincentive to the application in practice of such models. With the ready availability of inexpensive computer power, and its widespread use in design, we decided to construct the framework in the form of an interactive Web-based design support system (Steele *et al.*, 2000b). Computer-aided tools to support the design process are increasingly common, but often their rigidity has failed to support the informal nature of conceptual design work (Gardiner and Theobald, 1999). Most have been found usable only in the latter stages of design, and even there they provide only limited support (Baya and Leifer, 1996). Additionally, existing IT tools rarely help to link the various design disciplines. So far as we are aware, there is no commercial computer-based design tool to support the conceptual phase of construction design (Jensen, 1999), and our prototype Web-based design system was devised to demonstrate the principles such a system should embody.

Observations made during the experimental workshops revealed that the framework activities, although representative of the conceptual design phase, were not followed sequentially but instead were fragmented, with many loops and iterations. However, interdependencies between activities within and across particular phases of the design activity were also apparent. This inherent need for flexibility and non-linearity was seen as being prerequisite to delivering an adaptable yet responsive system that would be capable of aiding the design process without imposing a rigid procedure.

## System characteristics

The prototype exists as an interactive Internet-based design support tool, written in HTML. It may be entered at any one of the five main phases (interpret, develop, diverge, transform, converge) – it is not necessary to enter, as it were, 'at the beginning'. The team is asked a challenging question – for example, 'Is the team aware of the client's priorities among competing objectives?'. These questions are intended to discover whether the team is confident of having completed a phase and is ready to move to another. The questions have been devised to stimulate debate between the team members, as it is through negotiation between the various design disciplines that the design process is driven (Cuff, 1982).

Whatever the response, the team is always at liberty to move to any other activity, but by default the system will guide them to the next sequential activity. The pattern of progression through the system can be recorded for later review. If the team is not confident that it has addressed an issue adequately, the system offers some guidance. 'Team Thinking Tools', based on existing design methods such as brainstorming, synectics (forced analogies), and ranking and weighting methods are available online to help the team broaden the solution space, set priorities among competing objectives, or evaluate design alternatives.

In addition to offering guidance and tools, the system offers assistance with team management issues, such as the negotiation of roles, rights and responsibilities, effective communication, team dynamics, and social interaction.

Finally, the system offers the users the opportunity to record decisions as they are made during the course of designing. This enables the team to record who took a decision, who else contributed, what information it was based on, and the justification or reasoning behind it. The capture of this information, together with its storage and retrieval, has the potential to be valuable not only during the remainder of the project, but also in future projects. Such information could facilitate design reviews, provide an audit trail of decision making and contribute to the management of knowledge with the organizations represented in the team.

## Trialing the system in industry

The prototype was demonstrated to all the industrial collaborators in their offices and was revised in the light of feedback. Subsequently, a live trial of the revised prototype was organized with the project collaborators during a workshop to design part of a £100m-plus airport terminal building. The workshop focused on the design of one of the project packages – a flight pier – and it was arranged to mark the transition between the feasibility study and conceptual design (using the BAA Project Process (1995) phase description). The Web-based system was projected onto a screen at the front of the room, to be visible to the whole team, and its use was facilitated by the first-named author. The team's reactions to the system were recorded during the workshop, and there was also a round table discussion with the designers.

The system's availability on the Web was welcomed by the team members, who argued that conceptual design activity would be enhanced and more efficient if this type of guidance system, incorporating appropriate links to Web-based design information, were used in practice. They suggested it should be used in conjunction with a project document management system or an organizational database, such that decisions made at the early stages of the design process could be based on as much available data as possible.

It was the team's view that the system would have to be facilitated. This was not only because the designers would not have a full working knowledge of the system, but also because they wanted to focus on designing. The team agreed that the facilitator's mapping of the design progression of the team and his or her raising of the issues within the framework at the appropriate point in the process were both valuable. These actions concentrated the efforts of the team and made them reflect on their design activity. Making the team aware of the process assisted them in consciously negotiating among themselves the next step to be taken.

Within the workshop the team's use of the system focused on the evolving pattern of design progression. In the post-design discussion it was suggested that, in practice, the design manager could facilitate the use of the system. If the design manager monitored the design progression of the team using the system it would only be necessary to offer system advice when it was actually required (it could be used as a contingency tool) rather than continuously throughout the design process, irrespective of whether the team were functioning inefficiently or not.

The design team thought that the pattern of design progression that was produced as the team advanced was an extremely useful means of understanding the design process. The team acknowledged that members are generally not inclined to stop designing in order to pause and reflect upon their design process. However, within the workshop the design team referred regularly to the evolving pattern of design progression as a means of contemplating their design activity, highlighting sticking points in progression, and deciding on the next activity to be undertaken.

The (potential for a) decision-recording mechanism was thought by the team to be highly beneficial if applied in practice, and some went so far as to suggest that it would be a key design tool even without the framework structure in place. However, in the discussion that ensued as a result of this possibility, the design team acknowledged that it would be very difficult, and possibly impractical, to record decisions without relating them to the design process. One problematic issue raised with regard to the decision-recording function was that the activity invades design time. Decision capture is recognized as being vitally important, as it allows a design team/designer to defend a concept, option or project once complete, and to explain why they did what they did. However, how to develop a mechanism for doing this during design activity in a non-invasive manner should be the real question. It was suggested that if the decisions and rationale can be captured and recorded quickly and simply (without expending design time and effort) then designers will accept it; if it takes too long and requires lots of extra work, they will not.

The designers agreed that the support system has the potential to allow the team to leave a process trail that incorporates the capture of a great deal of knowledge. However, it is apparent that for a design rationale or decision-recording mechanism to be accepted in practice it must be both non-invasive and simple to apply. In this respect it appears that any tool which deals with the capture of the design rationale must be integrated into the everyday practices of design. The recognized strategy for achieving this is to embody rationale capture in tools that are of immediate utility to designers (Fisher, 1995). This became one of the goals of our Web-based system, but although we have made some progress, it was not possible to implement and test it within this research programme.

The results of applying the prototype system on a live design project workshop and the subsequent feedback from the project team on the potential of a Web-based version of the framework were highly encouraging. In general, the designers were very positive about the potential of the framework as a Web-based system. All team members agreed that Web-based communication was going to be a major component of design collaboration in the very near future. As such, they felt that, with further development, the Web-based version of the framework would offer a valuable means of achieving it in practice.

## Closing remarks

The vibrancy and dynamism of the conceptual design phase of building projects often leads to the design activity being chaotic in appearance. Iterative, or cyclic, design progression is critical in developing nebulous concepts into viable proposals. Design is a learning activity and, owing to the complexity of contemporary building projects, it is only by moving ahead to improve knowledge, before taking a step back to re-address a problem with improved understanding, that the design process can progress (Steele *et al.*, 2000a). Today's design professionals are being urged to undertake early design activity in a more programmable, and thus manageable, fashion. Yet without a shared interdisciplinary understanding of what conceptual design involves for each of the parties, nor a clear image of iteration as a key part of the process, there can be a lack of synchronization in interdisciplinary thinking, and, in consequence, fragmentation and adversity in design teams.

By modelling design activity it has been possible to explore some of the complexities of interdisciplinary conceptual design. This has led to the development and trialing of a generic framework of design phases and activities, which in turn has allowed a simple graphical means of recording and displaying patterns of design progression to be devised (Steele *et al.*, 2000a). The models produced have been used to study and analyze patterns of iterative working, and have led to a clarification of interdisciplinary conceptual design practice.

A Web-based design system has been developed from the paper-based framework. This accords well with the richly iterative and often non-linear process which design typically follows and is intended to encourage creativity without imposing a rigid procedure. The tool offers alternative routes through conceptual design, and contains 'Team Thinking Tools' to help designers widen the solution space, set priorities and evaluate options. In addition, it promotes effective teamwork practices to help teams deal with social interactions. At the user's option, the system has the potential to capture, store and retrieve decisions made and the reasoning behind them. Overall we believe that the research contributes both to improving interdisciplinary understanding of the early design stages, and also to a climate of opinion that will help good design to flourish.

## Acknowledgements

This chapter is based on the outcome of a research project called 'Mapping the Design Process at the early stages of construction projects'. It was supported by the Engineering and Physical Sciences Research Council. The industrial collaborators were: AMEC Design; BAA, WSP, Pascall & Watson, Hotchkiss Ductwork Ltd, and Matthew Hall. Loughborough University also contributed to the project.

## References

Archer, B. L. (1984) Systematic method for designers. In *Developments in design methodology* (ed. N. Cross). John Wiley & Sons, Chichester.

Baya, V. and Leifer, L. J. (1996) Understanding Information Management in Conceptual Design. In *Analysing Design Activity* (eds. N. Cross, H. Christiaans and K. Dorst). John Wiley & Sons, Chichester, pp. 151–168.

Bucciarelli, L. L. (1988) An ethnographic perspective on engineering design. *Design Studies*, **9**, 159–168.

Christiaans, H. and Dorst, K. (1992) An empirical study into design thinking. In *Research in Design Thinking* (eds. N. Cross *et al.*). Delft University.

Cross, N. and Clayburn-Cross, A. (1996) Observations of teamwork and social processes in design. In *Analysing Design Activity* (eds. N. Cross, H. Christiaans and K. Dorst). John Wiley & Sons, Chichester, pp. 291–317.

Cuff, D. C. (1982) Negotiating architecture: a study of architects and clients in design practice. *Ph.D. Dissertation*, University of California, Berkeley.

Gardiner and Theobald Management Services (1999). In *Columbus Prerelease Seminar Document*. Ove Arup and Associates/Oasys, December.

Hales, C. (1987) Analysis of the engineering design process in an industrial context. *Ph.D. Thesis*. Cambridge University Press, Cambridge.

Hedges, I. W., Hanby, V. I. and Murray, M. A. P. (1993) Logic modelling of the building design process. *Proceedings of the 1993 CIBSE Conference*, Manchester, pp. 28–39.

Jensen, T. (1999) Function modelling in a design support system: contribution to a designer's workbench. *Ph.D. Thesis*, Department of Control and Engineering Design, Technical University of Denmark.

Macmillan, S., Steele, J., Austin, S., Kirby, P. Spence, R. (1999) Mapping the early stages of the design process – a comparison between engineering and construction. *Proceedings of the 12th International Conference on Engineering Design 1999*, Munich, Germany, pp. 1175–1178.

Macmillan, S., Steele, J., Austin, S., Kirby, P. and Spence, R. (2001). Development and verification of a framework for conceptual design. *Design Studies*, **22**(2), 169–191.

Minneman, S. L. (1991) The social construction of a technical reality: empirical studies of group engineering design practice. *Ph.D. Dissertation*, Stanford University.

Peng, C. (1999) Flexible generic frameworks and multidisciplinary synthesis of built form. *Design Studies*, **20**(6), 537–551.

Radcliffe, D. F. (1997) Event scales and social dimensions in design practice. In *Proceedings of the Darmstadt Symposium: 'Designers – the key to successful product development'*, 3–5 December, Darmstadt, Germany.

Radcliffe, D. F. and Lee, T. Y. (1990) Models of visual thinking in novice designers. *Design Theory and Methodology – ASME, DE*, **27**, 145–152.

Steele, J., Austin, S., Macmillan, S., Kirby, P. and Spence, R. (1999) Interdisciplinary interaction during concept design. *Proceedings of the 15th Annual Conference of the Association of Researchers in Construction Management 1999*, Liverpool, pp. 297–305.

Steele, J., Austin, S., Macmillan, S., Kirby, P. and Spence, R. (2000a). One step forward and three back: a study of the patterns of interdisciplinary conceptual design. *Proceedings of the Design Research Conference*, Coventry, September.

Steele, J., Austin, S., Macmillan, S., Kirby, P. and Spence, R. (2000b). An internet-based system to support interdisciplinary and inter-organizational collaborative conceptual design. *Proceedings of the Chartered*

*Institute of Building Services Engineers (CIBSE) National Conference 2000*, Dublin.

Steele, J. L. (2000) The interdisciplinary conceptual design of buildings. *Ph.D. Thesis*. Department of Civil and Building Engineering, Loughborough University.

Tang, J. and Leifer, L. J. (1991) An observational methodology for studying group design activity. *Research in Engineering Design*, **2**, 209–219.

Valkenburg, R. and Dorst, K. (1998) The reflective practice of design teams. *Design Studies*, **19**(3), 249–271.

# Index